Children of Light

Children of Light

The Astonishing Properties of Sunlight that Make Us Possible

Michael Denton

Seattle Discovery Institute Press 2018

Description

We associate light with the radiant beams that make the world visible to us. But the visible spectrum is only a tiny percentage of an electromagnetic spectrum that extends unimaginably far in both directions. And, as biologist Michael Denton carefully documents, that tiny band of visual light is crucial to life on Earth.

In *Children of Light*, Denton elucidates the miraculous convergence of properties on the tiny band we call the visible spectrum that has allowed intelligent life to flourish on Earth. Follow the journey of light as it beams down from our Sun, through the protective blanket of our atmosphere, to the Earth. Once here, it powers photosynthesis and unlocks the oxygen needed for life. It allows the high-acuity vision that led us to civilization and technology. Light is just one more part of the epic story of our fine-tuned universe, fit for us to flourish here and come to understand it.

Copyright Notice

Copyright © 2018 by Discovery Institute. All Rights Reserved.

Library Cataloging Data

Children of Light: The Astonishing Properties of Sunlight that Make Us Possible by Michael Denton

166 pages, 6 x 9 x 0.36 in. & 0.5 lb, 229 x 152 x 9 mm & 232 g

BISAC: SCI004000 Science/Astronomy

BISAC: SCI008000 Science/Life Sciences/Biology

BISAC: SCI015000 Science/Cosmology

BISAC: SCI005000 Science/Physics/Astrophysics

BISAC: SCI053000 Science/Physics/Optics & Light

ISBN-13: 978-1-936599-59-2 (paperback), 978-1-936599-60-8 (Kindle), 978-1-936599-61-5 (EPUB)

Publisher Information

Discovery Institute Press, 208 Columbia Street, Seattle, WA 98104

Internet: http://www. discoveryinstitutepress.org/

Published in the United States of America on acid-free paper.

First Edition: October 2018.

Learn More about Michael Denton and the Privileged Species Series

Explore additional books and view documentaries related to the content of this book.

www.PrivilegedSpecies.com

Contents

Preface

THIS BOOK DOCUMENTS THE MANY REMARKABLE WAYS IN WHICH the light of the Sun and the properties of the Earth's atmosphere are both supremely fit for photosynthesis, the most important chemical reaction on planet Earth for aerobic beings of our biological design. It also describes the unique fitness of light for high-acuity vision in biological beings of our anatomical design and size. It complements the two previous books *Fire-Maker* and *The Wonder of Water* by providing additional compelling evidence that nature is not merely biocentric, but can be considered in a very real sense to be also *anthropocentric*.

Because the book is intended to be a stand-alone publication, and also intended to cover, fairly comprehensively, specific phenomena that demonstrate the unique fitness of nature for our type of being, it was necessary to go over some of the same ground already dealt with in some of my previous books, including *Fire-Maker, The Wonder of Water,* and *Nature's Destiny*. Consequently, some sections of the text have been abstracted from previous publications.

The fitness of nature for terrestrial plant life, which I dealt with extensively in *The Wonder of Water,* is reviewed again in Chapter 4. This is an appropriate revisit because only air-breathing organisms are able to extract sufficient oxygen to sustain their high metabolic rates and only a terrestrial environment is fit for fire-making and its consequence, the discovery of metallurgy, which initiated the journey to our present-day technological civilization. It is terrestrial plants and trees which enable both respiration, by providing the reduced carbon nutrients for oxidation in the body, and combustion, by providing the woody fuels for fire-making.

Chapter 5, "Fitness for Vision," revisits the evidence for the fitness of light for high-acuity vision presented in Chapter 3 of *Nature's Destiny,* with significant additional insights and updates. Additionally, Chapter 3 contains a short section on silicate weathering which has been abstracted from *The Wonder of Water,* Chapter 3.

While this book discusses the fitness of sunlight for photosynthesis and the *provision* of oxygen and the reduced carbon fuels we advanced aerobes burn to generate metabolic energy, it does not deal with nature's fitness for the *utilization* of oxygen by organisms of our physiological design. Nature's fitness for the use of oxygen depends on another very remarkable suite of elements of fitness. Consequently, there are important elements of fitness in nature for aerobic life *not* reviewed here, including the fitness of transitional metals such as iron and copper to handle and tame oxygen. For instance, iron plays a key role in O_2 transport in hemoglobin, and the transitional metals form the electron-conducting wires in electron transport chains which are involved in generating the proton gradients involved in the synthesis of ATP, the key provider of chemical energy in the cell.

Likewise, there is another whole suite of elements of fitness in nature that allow the uptake of sufficient oxygen to satisfy our high metabolic rate. But, again, that is a topic for another book.

I have tried to adhere to the highest level of scholarship, and I hope that all the very many facts alluded to are correct and properly cited. I have also endeavored to make the text as accessible to a non-technical reader as possible. In places the text is a bit heavy-going—in describing the chemical process of photosynthesis, for example. The reader may skip these sections and still follow the main thrust of the argument. I have also put some of the more technical and esoteric points in appendices or extensive footnotes so as not to detract from the flow of the argument.

FIGURE 1.1. Sunrise at Stonehenge (Wiltshire, England) on the summer solstice of June 21, 2005.

1. The Miracle of Sunlight

> Think of the Sun's heat on your upturned face on a cloudless summer's day; think how dangerous it is to gaze at the Sun directly. From 150 million kilometers away, we recognize its power. What would we feel on its seething self-luminous surface, or immersed in its heart of nuclear fire? The Sun warms us and feeds us and permits us to see. It fecundated the Earth. It is powerful beyond human experience. Birds greet the sunrise with an audible ecstasy. Even some one-celled organisms know to swim to the light. Our ancestors worshiped the Sun, and they were far from foolish.
>
> —Carl Sagan (1980), *Cosmos*[1]

Even after waiting for most of the night at Stonehenge, blanket wrapped around one's shoulders against the cold, and despite the crowded throng of druid worshippers and ecstatic new-agers, it is hard not to be enthralled by the magic of the Sun slowly rising above the horizon just to the left of the heel stone on the morning of the summer solstice. And, as Richard Cohen recalls, even on days when the site is deserted at either sunrise or sunset: "[I]t is magical, and one's imagination soars. Against the lateral sunlight, the stones gain in beauty and majesty."[2] No one who has visited this dramatic and most celebrated of all the world's solar monuments, built during the third millennium B.C. on what is now the Wiltshire Downs in southern England, can fail to sense the magic of the place.

Of course, Stonehenge is not the only solar monument. Thousands of constructions around the world are also aligned to the Sun, either on the solar solstices or on its position on the horizon on the spring and autumn equinoxes. Examples abound, including the Sphinx at Giza,[3] the bas-reliefs in the largest of all religious complexes on Earth, the Angkor Wat temple complex in Cambodia,[4] the solar stone circles of the North

American Indians,[5] and whole cities in pre-Columbian Mesoamerica, such as Teotihuacan,[6] and their associated pyramids and monuments.[7] And true to animist conceptions, the Sun himself was considered a living deity by many ancient cultures throughout the world, especially in Mesoamerica, among Indo-European peoples, and in ancient Egypt,[8] where the Sun god was Re (or Ra), the creator of life and nourisher of the Earth.[9] Many solar cultures identified the Sun-god's motion across the sky with the flight of a falcon or bird.[10] The Indo-Europeans portrayed him as guiding a chariot drawn by four flaming steeds.[11]

Today we neither worship nor build monuments to the Sun. It is true, however, that we are still enthralled by the Sun, such as when watching a spectacular sunset, as the red orb sedately sinks below the western horizon. Who is not enchanted in high mountain country by the Alpen glow, that faint red radiance which lingers on the high snow fields after sunset, like a spectral presence before the dark finally descends? Or who is not intrigued by the eerie twilight during a total eclipse? We, as much as our ancestors, still feel his radiant power at midday in tropical or even temperate climates, when animals shelter from his fierce glare and prolonged exposure causes sunburn. We still celebrate the coming of the spring and the re-greening of the land. And on rare occasions, such as the summer solstice at Stonehenge, we revert for a moment in reverie, and recapture our ancestors' sense of wonder at his return from the kingdom of the dark to the realm of the light.

But on the whole, we seldom give the Sun and his life-giving radiation a second thought. As Bob Berman commented, "Familiarity is the enemy of awe, and for the most part people walk the streets with no upward glance [or thought of the Sun]."[12] Yet if there is any one entity in the natural world worthy of awe, if not of worship, it must surely be the Sun! This is particularly true for us and other creatures whom William Broad calls "light eaters,"[13] denizens of the Earth's surface whose existence depends entirely on the radiant output of the Sun.

Without the warmth of the Sun, all the water on the surface of the Earth would be imprisoned in an eternal icy tomb. And without liquid water, there would be no oceans, no hydrological cycle providing water for terrestrial life, no climate machine moderating the Earth's climate, no plate tectonics to refresh the chemistry of the hydrosphere, and no life like our own, based in an aqueous matrix. While water is supremely fit for its various vital roles on Earth, as reviewed in *The Wonder of Water*, it is only the warmth of the Sun, beamed to Earth over billions of years, which has released water's unique life-giving talents.

Take away the Sun and after a short time the Earth's surface would be a lifeless frozen waste only a few degrees above absolute zero—colder, far colder, than the center of the Antarctic or even than the frozen sands of Mars. Every atom and molecule on the Earth's surface would be immobilized and all chemistry would cease. Even the partial blocking out of the Sun's radiance for a few years at the end of the Cretaceous following the Chicxulub meteor strike caused a mass extinction event, bringing to an end the age of the dinosaurs.[14]

Most of us are aware that it is the light of the Sun via photosynthesis that bestows life upon ourselves, and all advanced organisms on Earth, by providing the basic food-stuffs upon which we depend as well as the precious oxygen, which we use to oxidize those foodstuffs in the body, providing the energy on which our advanced metabolism and active lifestyle depends. Less widely known, however, is the existence of an extraordinary suite of coincidences in the nature of things which render the Earth's surface a supremely fit habitat for advanced carbon-based life forms like ourselves—coincidences that are, on any consideration, ludicrously improbable.

To begin with, the electromagnetic radiation emitted by the Sun (and that of most other stars) is almost entirely light and heat (or infrared), which have precisely the characteristics needed for life, especially advanced life, to thrive on the Earth's surface. *Light* is required for photosynthesis and *heat* is required to raise the Earth's temperature to well above freezing and preserve liquid water on Earth.

It is only because of the precise absorption characteristics of Earth's atmospheric gases that most of the light radiation emitted by the Sun reaches the Earth's surface where it drives the chemical process of photosynthesis upon which we "light eaters" ultimately depend. And the same atmospheric gases which let the light through for photosynthesis absorb a portion of the infrared (IR) radiation, which warms the Earth and preserves water as a liquid on the Earth's surface. Adding to the miracle, both the atmospheric gases and liquid water, the matrix of carbon-based life, not only let through the right light but strongly absorb all the dangerous types of radiant energy on either side of the visual and infrared regions of the electromagnetic spectrum, a vital property without which no advanced life forms would grace the surface of the Earth.

In addition to being perfectly fit for photosynthesis and hence for our kind of oxygen-utilizing advanced life, sunlight is also just right for high-acuity vision, which depends on another set of extraordinary coincidences in the characteristics of visual light. And sunlight is just right not only for *any* type of high-acuity visual device or eye, but just right in terms of its wavelength for beings of our size and upright android design.

What is so significant about the fitness of the Sun's light for photosynthesis and for high-acuity vision is that these are elements of natural fitness *exclusively for our type of life—for beings possessing the gift of sight, breathing oxygen (aerobic), and inhabiting the terrestrial surface of a planet like the Earth.* The wondrous fitness of the Sun's radiation and the many other elements of fine-tuning in nature for photosynthesis and hence for advanced aerobes like ourselves are quite irrelevant to William Broad's "alien horde"[15] that thrive in total darkness, either in the crustal rocks or near the hydrothermal vents in the abyssal ocean depths, feeding on minerals leached from the rocks, surviving on geothermal warmth, and generating energy from chemical reactions which do not involve oxygen.

In documenting the supreme fitness of sunlight for life on Earth and especially for *our* kind of life, this book graphically illustrates that the ancient Sun-worshippers who gathered millennia ago on June 21 at Stonehenge, although lacking any scientific knowledge of the fitness of the Sun or atmosphere, were, as Sagan confessed, "far from foolish." Indeed, the story of the utility of sunlight and the vastly improbable coincidences which make possible our specific type of being—for the precious oxygen we crave which fires our metabolism and for the great gift of vision itself—is a tale far more extraordinary than any told by any of the ancient Sun-worshippers or conceived of in any of the most esoteric visions of the gurus and seers of the past, and certainly far more extraordinary than anything conceived of by the ecstatic hippies on the Wiltshire Downs on the morning of the summer solstice.

FIGURE 2.1. The life-giving Sun.

2. The Light of Life

> It may form an interesting intellectual exercise to imagine ways in which life might arise, and having arisen might maintain itself, on a dark planet; but I doubt very much that this has ever happened, or that it can happen.
>
> —George Wald (1959), *Scientific American*[1]

As is common knowledge, the light which drives the process of photosynthesis and provides us "light eaters"[2] with the oxygen we breathe and the foodstuffs we eat is provided by the Sun. And it is the heat radiated by the Sun which renders the Earth's surface habitable, by raising the temperature of the atmosphere well above freezing to an average global balmy temperature of about 15°C.[3]

The spectrum of radiation emitted by any physical body (including a star like the Sun; see Figure 2.2) is largely determined by its surface temperature.[4] At a surface temperature of close to 6,000°C, the temperature of the Sun's surface,[5] most of the radiant emission takes the form of visual light and heat.[6] At temperatures lower than the Sun's, a greater proportion of the radiant output is in the infrared, or heat region, with less in the visual region. At temperatures higher than the Sun's, a higher proportion is in the UV region, which is harmful to life. However, stars that radiate mainly dangerous UV or more energetic radiation represent only a tiny minority of all stars in the cosmos.[7] The majority of stars have surface temperatures close to or lower than 6,000°C, and emit nearly all their radiation, like the Sun, as light and heat.[8] Consequently, in terms of its radiance the Sun is not exceptional, but rather, as described by Carl Sagan, "an ordinary, even a mediocre star."[9] Ordinary or not, the Sun provides precisely the kind of light that is helpful for life.

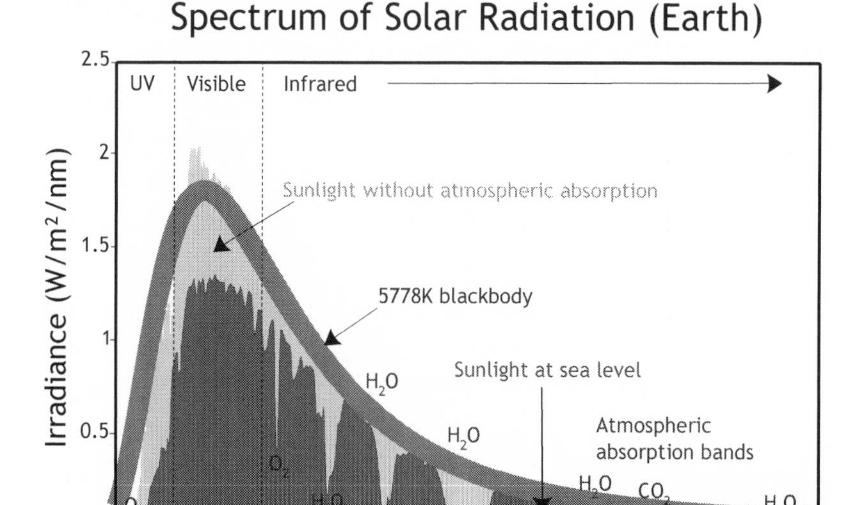

FIGURE 2.2. The solar spectrum. Note that the solar spectrum extends somewhat further into the IR region than shown. Other depictions show the spectrum as extending to about 5,000 nanometers or five microns.

The light of the Sun is fit for advanced life on a planetary surface in another way. The life of stars like our Sun is measured in billions of years.[10] The Sun will continue to put out radiation in the visual and IR bands—the right light for life—for another four billion years. This means that life on the surface of the Earth (and of any rocky planet in the habitable zone circulating a typical star) is ensured an ideal source of energy over enormous time spans, which, while beyond our ordinary experience, are necessary if life is to develop and thrive on the surface of a planet over periods of millions of years. Moreover, there is also very little day-to-day, year-to-year change in the solar output—another crucial element of fitness without which the Earth's climate would be unstable and unsuited for the thriving of complex, advanced life.

A further intriguing element of fitness is that stars of the same mass as the Sun have *both* lifetimes of many billions of years (ten billion in the case of the Sun) *and* the right temperature, approximately 6,000°C, to emit the "right" radiation in the visual and IR region of the spectrum.[11] Even stars only three times the Sun's mass have vastly shorter life spans. A star of three times the mass of the Sun has a life span of 370 million years,[12] less than the time span of terrestrial life on Earth. On the other hand, stars of less mass than our Sun have longer lifetimes[13] but, being cooler,[14] put out less visual light as a proportion of their total radiance.[15]

Although it is well understood that the fitness of the radiant output of the Sun for life on Earth is due to its surface temperature being close to 6,000°C, what is not appreciated is that the fitness of the Sun's radiation for life depends on a number of extraordinarily improbable coincidences in the nature of things. These can only be appreciated by considering the electromagnetic spectrum and the interaction between electromagnetic radiation and matter.

The Electromagnetic Spectrum

The electromagnetic spectrum (EMS) consists of many different types of radiation (see Figure 2.3). The two most familiar forms are light and heat, such as our Sun primarily emits. (Heat is infrared or IR radiation, which we experience as warmth when the Sun shines on the skin.) But there are many other kinds, such as radio waves, microwaves, ultraviolet, X-rays, and gamma rays. All the various different types of electromagnetic radiation, including light and heat (or infrared), flow through space in the form of energy waves analogous to ripples on the surface of a pond. And just as the waves on the surface of a pond may have different wavelengths—small ripples may be only a few millimeters from crest to crest, while large waves might be several centimeters from crest to

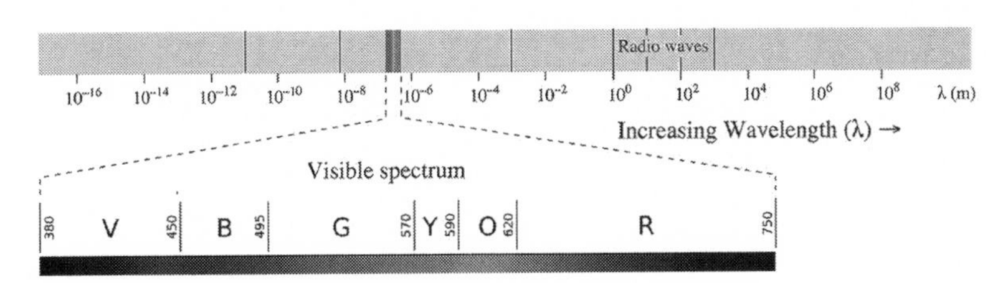

Figure 2.3. The electromagnetic spectrum. λ (lambda) = wavelength

crest—so similarly the wavelength of the various types of electromagnetic radiation also varies, but over a vastly greater range.

Electromagnetic radiation, from the longest to the shortest waves, pervades all of nature, as the article on EM radiation in the *Encyclopaedia Britannica* points out:

> Close to 0.01 percent of the mass/energy of the entire universe occurs in the form of electromagnetic radiation. All human life is immersed in it, and modern communications technology and medical services are particularly dependent on one or another of its forms... food is heated in microwave ovens, airplanes are guided by radar waves, television sets receive electromagnetic waves transmitted by broadcasting stations, and infrared waves from heaters provide warmth. Infrared waves also are given off and received by automatic self-focusing cameras that electronically measure and set the correct distance to the object to be photographed. As soon as the Sun sets, incandescent or fluorescent lights are turned on to provide artificial illumination, and cities glow brightly with the colourful fluorescent and neon lamps... Less familiar are gamma rays, which come from nuclear reactions and radioactive decay and are part of the harmful high energy radiation of radioactive materials and nuclear weapons.[16]

The reason that the effects and uses of the types of radiation in different regions of the EM spectrum are so varied is that wavelengths of different length—microwave, IR (heat), visual, ultraviolet, and X-rays—interact with matter in very diverse ways.[17]

The total range of wavelengths in the EM spectrum is inconceivably vast. Some extremely low-frequency radio waves may be a hundred thousand kilometers from crest to crest, while some higher-energy gamma waves may be as little as 10^{-17} meters across (only a fraction of the diameter of an atomic nucleus). Even within this selected segment of the entire spectrum the wavelengths vary by an unimaginably large factor of 10^{25} or 10,000,000,000,000,000,000,000,000.

Some idea of the immensity this figure represents can be grasped by the comparison I offered in *Nature's Destiny*:

> [T]he number of seconds since the formation of the earth 4 billion years ago, is *only* about 10^{17}. To count 10^{25} seconds we would have to keep counting every day and night through a period of time equal to *100 million times the age of the earth!*[18]

Put another way, if we were to build a pile of 10^{25} playing cards, we would end up with a stack stretching from Earth to beyond the Andromeda Galaxy.[19]

The Right Light

As mentioned earlier, EM radiation interacts with matter in many different ways[20] and exhibits: "[A] multitude of phenomena as it interacts with... atoms, molecules, and larger masses of matter."[21] But the only type of radiation (or type of photons) in the entire electromagnetic spectrum that enables photochemistry, including photoreception and photosynthesis, having just the right energy level to raise electrons to higher energy levels (see Figure 2.4) and activate chemicals for chemical reaction, *is that within the visual region.*[22] As Nobel Laureate George Wald, who worked for many years on the biochemistry of vision, pointed out:

> Almost all ordinary... [chemical] reactions involve energies of activation between 15 and 65 kilogram calories (kilocalories) per mole.

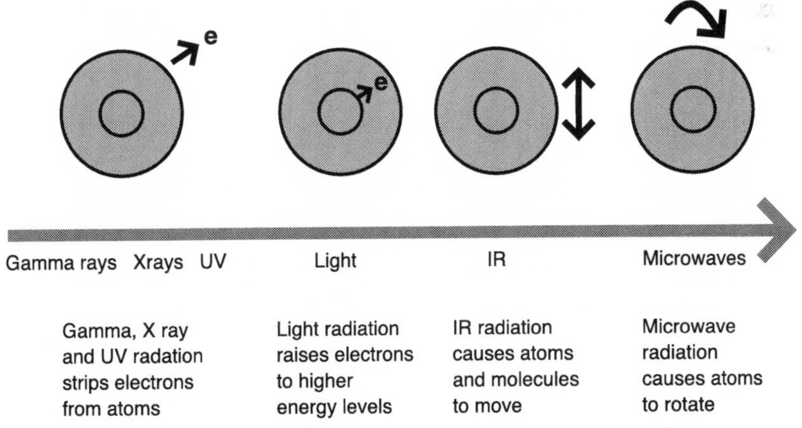

Figure 2.4. Interaction of EM radiation with matter. Less energetic radiation of wavelengths longer than UV is non-ionizing; more energetic radiation of wavelengths shorter than visual (e.g., UV, X-rays) is ionizing and can strip electrons from atoms and molecules, causing chemical damage.

> This is equivalent energetically to radiation of wavelengths between 1,900 and 440 millimicrons. The energies required to break single covalent bonds—a process that... can be a potent means of chemical activation—almost all fall between 40 and 90 kilocalories per mole, corresponding to radiation of wavelengths 710 to 320 millimicrons. Finally, there is the excitation of valence electrons to higher orbital levels that activates the reactions [such as occurs when photons are absorbed by chlorophyll] classified under the heading of photochemistry; this ordinarily involves energies of about 20 to 100 kilocalories per mole, corresponding to the absorption of light of wavelengths 1,430 to 280 millimicrons. Thus, however one approaches the activation of molecules for chemical reactions, one enters into a range of wavelengths that coincides approximately with the photobiological domain.[23]

Radiant energy outside the visual region is unfit for photochemistry.[24] In the IR and microwave region of the electromagnetic spectrum (wavelengths greater than 0.80 microns) the photons are too weak—causing molecular vibrations and rotations but incapable of imparting to atoms and molecules sufficient energy to activate molecules for chemical reaction. (Note that at temperatures higher than the ambient temperature range in which life's chemistry occurs, heat can cause molecular collision of sufficient force to activate atoms and molecules for chemical reaction.) On the other hand, photons in the far UV, X-ray, or gamma regions (wavelengths less than 0.3 microns)—termed ionizing—are too energetic, imparting so much energy that they cause major disruptive changes, stripping electrons from atoms and molecules, ionizing them, breaking chemical bonds, and leading for example to mutations in DNA. The disruptive effect of radiation in the UV and beyond is why radiation in these regions is inimical to life.

As Wald commented:

> Radiations below 300 millimicrons... are incompatible with the orderly existence of such large, highly organized molecules as proteins and nucleic acids. Both types of molecule consist of long chains of units bound together by primary valences [ordinary chemical bonds]. Both types of molecule, however, are held in the delicate and

> specific configurations upon which their functions in the cell depend by the relatively weak forces of hydrogen-bonding and van der Waals attraction. These forces, though individually weak, are cumulative. They hold a molecule together in a specific arrangement, like zippers. Radiation of wavelengths shorter than 300 millimicrons [0.3 microns] unzips them, opening up long sections of attachment, and permitting the orderly arrangement to become random and chaotic. Hence such radiations denature proteins and depolymerize nucleic acids with disastrous consequences for the cell. For this reason about 300 millimicrons represents the lower limit of radiation capable of promoting photoreactions, yet compatible with life.[25]

In other words, light in the visual region of the EM spectrum is just right for photochemistry. Within this Goldilocks region, the light is not so energetic as to cause chemical disruption of organic matter, but it is energetic enough to gently activate organic molecules for chemical reaction. In other words "just right." No other EM radiation will do! As Wald points out, it is not that life adapted to the right light but *that the right light is the only light that provides the correct energy levels for photochemistry*:

> There cannot be a planet on which photosynthesis or vision occurs in the far infrared or far ultraviolet, because these radiations are not appropriate to perform these functions. It is not the range of available radiation that sets the photobiological domain, but rather the availability of the proper range of wavelengths that decides whether living organisms can develop and light can act upon them in useful ways.[26]

This crucial visual band that has the right energy levels for photochemistry occupies only a very small part of the EM spectrum. In his article cited above, Wald alludes to the narrowness of the Goldilocks region of the EM spectrum:

> [L]ight... comprises only a narrow band in the spectrum of radiant energy that pervades the universe. From gamma rays, which may be only one ten billionth of a centimeter long... to radio waves, which may be miles in length. The portion of this spectrum that is visible to

man is mainly contained between the wavelengths 380 to 760 millimicrons.[27]

But such a relatively bland description, while true, fails to highlight just how extraordinarily tiny is the fraction of the total EM spectrum occupied by the visual region. Just how narrow is the Goldilocks band of the "right" light?

Most depictions of the EM spectrum show the wavelengths along a logarithmic scale and this inevitably conveys the impression that the visual region occupies a significant portion of the spectrum (see Figure 2.3 above). But such depictions are quite misleading. A more accurate depiction would represent the visual region as a thin vertical line between the UV and infrared regions of the spectrum (as shown in Figure 2.5). But even the thinnest of lines drawn on a diagram cannot capture the exceedingly tiny region represented by the visual band. To use the analogies from earlier in this chapter, the "right light" would be only a few seconds in a time-span one hundred million times longer than the age of the Earth, or a few playing cards in a stack stretching beyond the galaxy of Andromeda—a fraction so small as to be beyond ordinary human comprehension.

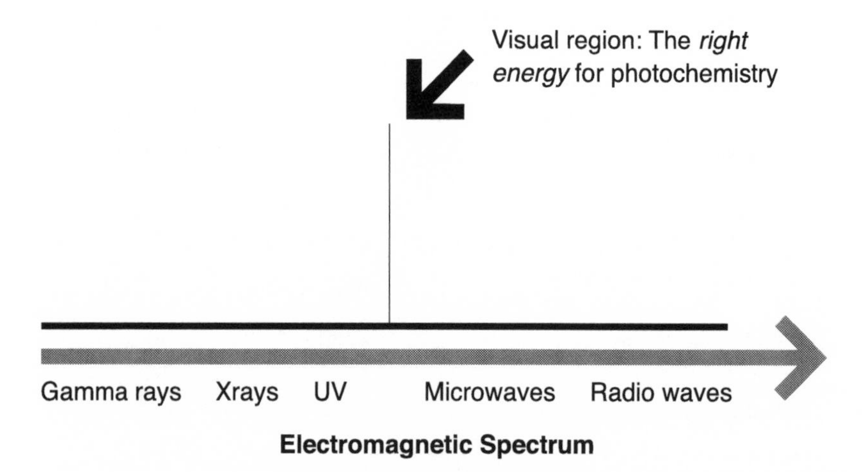

Figure 2.5. The visual region.

The Right Heat

In addition to light, the Sun also provides us with heat which keeps the surface waters from freezing solid, an essential element of fitness for carbon-based life forms, instantiated as they are in an aqueous matrix. Even at a temperature as relatively high as -20°C, vitrification of cell water occurs, immobilizing the cell's molecular constituents and causing all metabolic activity to cease.

Without the Sun to warm the Earth's atmosphere and hydrosphere, the entire surface would be a frozen wilderness far colder than the center of the Antarctic, close in temperature to absolute zero. Not only would all water be frozen, but the atmospheric gases would also be frozen solid, as they are on Neptune's moon Triton, a chilly world at –236°C and covered in a thin layer of nitrogen ice.[28] The only other source of heat on rocky planets like the Earth is internal geothermal heat, which provides warmth for the weird and wonderful biochemical denizens of the absolute dark in the depths of the sea round mid-ocean ridges (see Appendix 1). But the amount of geothermal heat that reaches the Earth's surface is only a small fraction of the heat provided by the IR radiation delivered by the Sun.

Heat is essential for another reason. In addition to the average ambient temperature remaining above the freezing point of water, for chemical reactions to occur, atoms and molecules must not only be activated or excited to higher energy levels but must also collide with each other to forge new chemical bonds. Only if molecules come into contact can a new chemical bond between two molecules be forged. Because it is heat which imparts motion to molecules, *heat is an essential ingredient* for any chemical reaction to occur.

But not too much! As temperatures rise, the speed and the force of molecular collisions increases and the likelihood that collisions may impart sufficient energy to overcome energy barriers and activate atoms and molecules for chemical reaction also increases. And this is the reason why, at temperatures much above the ambient range, controlled biochemistry becomes increasingly impossible. Uncontrolled chemical reactions enabled by heat-induced molecular collisions predominate. However, in the ambient temperature range, molecules seldom collide with sufficient energy to activate molecules for chemical reaction. This is why the familiar material forms of the animate (living things) and inanimate (rocks, buildings, cars, etc.) domains are relatively stable at ambient temperatures. Living things whose chemistry occurs within the ambient temperature range use energy-rich molecules like ATP to activate bio-molecules for chemical reactions and enzymes to overcome energy barriers to drive the specific reactions essential to life.

As the temperature drops, molecular motions become slower and slower, causing chemical reaction times to increase. At absolute zero, –273°C, all molecular motions cease and chemical reactions are no longer possible. The effect of temperature on the rates of chemical reaction is quite dramatic: for each 10°C fall in temperature, the rate of reaction declines by a factor of two.[29] It follows from this relationship that a fall in temperature of 100°C will slow chemical reactions nearly 1,000 times! Even a decrease in temperature of far less than 100°C causes a quite dramatic slowing of reaction times. Reactions occurring in the human body at 38°C would take place sixteen times slower at 0°C and sixty-four times slower at –20°C.[30] As Robert E. D. Clark pointed out:

> Experiments have shown that practically all reactions become vanishingly slow if the temperature is lowered to –50 or –100°C. At liquid air temperature [–200 C]—still very high compared with most of the matter in space—only a few reactions take place at all and these all involve the exceedingly active element fluorine in its free state.[31]

In passing, it is intriguing to note that we are fortunate indeed that there exists a temperature range in which the gentle, controlled chemistry of life can occur—where the heat is sufficient for molecules to gently touch each other and allow bonding to occur but not sufficient to overcome energy barriers resulting in uncontrolled reactions. This is the temperature range in which life exists on Earth, approximately between

–20°C[32] and 120°C,[33] which is the upper limit determined by the stability of weak and strong chemical bonds utilized in biochemical systems (see *The Wonder of Water*, Chapter 7, for further discussion).

And as I commented in *The Wonder of Water*: "Although this [temperature range] may appear large from our perspective, it represents an unimaginably tiny fraction of the total range of all temperatures in the cosmos."[34] Temperatures in the cosmos range from 10^{32} degrees C[35] (ten followed by thirty-one zeros), which was the temperature of the universe shortly after the Big Bang, to very close to absolute zero, -273°C. The temperature inside some of the hottest stars is several thousand million degrees,[36] and even inside our own Sun, which is not a particularly hot star, the temperature is in the order of 15,000,000°C.[37] So, the right temperature range for biochemistry is an inconceivably small fraction of the total cosmic temperature range.

Not only does this Goldilocks temperature range allow life's controlled chemistry to be carried out; it is almost exactly the same temperature range in which water, the ideal and perhaps only fluid matrix available for carbon-based life, exists as a liquid in the ambient conditions on the Earth's surface. In the *Wonder of Water*, I called this "The Prime Coincidence"[38] upon which the existence of carbon-based life in the universe is critically dependent.

The essential heat that prevents the Earth's hydrosphere from freezing solid and that animates matter for chemical reactions is provided by electromagnetic energy in another region of the EM spectrum—the IR region, or more specifically the near infrared. This region lies adjacent to the visual band, between it and the far infrared and microwave regions, or between about 0.8 microns and 14 microns.[39] This is the only region of the EM spectrum which can provide safe heat to warm the Earth, preventing it from freezing, providing sufficient kinetic energy to move molecules and promote chemical reactions but not enough to cause uncontrolled chemistry.

The only other type of EM radiation which might also provide heat to warm the Earth is in the far IR and microwave regions. But heating

Earth with microwave radiation rather than IR would likely have serious, deleterious biological consequences.[40]

While the near IR radiation band is somewhat longer than the visual band, it still represents an inconceivably tiny fraction of the EM spectrum—only a few cards in that deck that extends beyond Andromeda. This means that life on Earth, more specifically our type of life, depends not only on *one* inconceivably narrow band in the EM spectrum but on *two*: The visual, providing the right activation energy for photochemical reactions including those of photosynthesis, and the near IR, providing the right heat energy to prevent Earth's water from freezing and to mobilize atoms and molecules, bringing them into gentle contact with each other for controlled chemical reaction. These two vital bands, adjacent in the EM spectrum, even taken together represent an inconceivably small fraction of the EM spectrum. We can represent them both as an imaginary line of inconceivable thinness in the immense range of the EM wavelengths (see Figure 2.6).

The Life-Giving Miracle

It is at this stage that this consideration of the fitness of sunlight for life on Earth and for photosynthesis enters what can only be described as the realm of miracle.

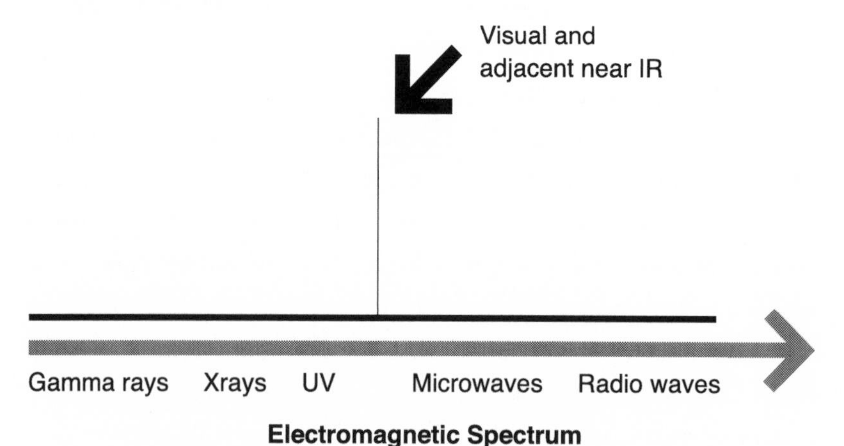

Figure 2.6. Visual band and adjacent near IR band.

The restriction of the two useful radiations—light and heat—into two impossibly tiny adjacent regions of the spectrum is remarkable in itself. But it is even more remarkable that (as shown above; see Fig 2.2) nearly all the electromagnetic radiation emitted from the surface of the Sun is concentrated in these same exceedingly narrow radiation bands, extending from the near ultraviolet through the visible region into the near infrared.[41] That the Sun should emit radiation in the only infinitely small region of the EM of utility to life is a truly extraordinary coincidence!

And because nearly all the energy output of the Sun is in the visual and near infrared region of the EM spectrum—the Goldilocks region—the Sun emits very little in the way of harmful ionizing radiation in the far UV, X-ray, and gamma regions, and very little radiation in the far infrared and microwave regions. The radiation output of the Sun *is exactly what is needed for life on Earth, all in the right or useful region of the EM spectrum, the light and heat regions, and virtually none in the dangerous regions on either side of the near infrared and near UV.*

Depiction of the Sun's radiation output in the visual and IR regions such as shown in Figure 2.2 seldom highlight just what a tiny fraction of the EM spectrum the Sun's output represents.

And of course, this is not only true of our Sun. As we saw earlier, most of the stars in the universe emit light within that same tiny region. Because stars put out such a high proportion of their radiation as light, light astronomy is still the most important branch of astronomy and the dark night sky twinkles with the light of stars (Figure 2.7). The Sun and the majority of stars put out just the right photons needed for photosynthesis in this impossibly tiny region of the EM spectrum. This means that the vast majority of planets circulating suns in the habitable zone throughout the cosmos will be similarly bathed in the right Goldilocks light.

The cosmos is, so to speak, flooded with the light of life! In the radiant output of the stars and the flooding of the cosmos with the right

FIGURE 2.7. The Milky Way in the night sky over Black Rock Desert, Nevada.

light, nature signifies *her* unique fitness for "light eating" organisms like ourselves.

Further, it is worth noting that the long life of stars and their ability to radiate continuously and reliably both light and IR for billions of years, which is a vital element of fitness for life on planets like the Earth, is crucially dependent on the mechanism of nuclear fusion which keeps stars alive, supplying them with copious amounts of energy during their long lifetimes. This mechanism—nuclear fusion—is itself dependent on the nuclei having just the right energy levels to allow key steps in the synthesis of the heavier elements, including the core elements necessary for life to proceed.[42] Nuclear fusion is also dependent on a phenomenon termed quantum tunneling (see below for further discussion) which enables the nuclei to fuse, by surmounting the protective energy barrier which would otherwise prevent nuclei from coming into close contact.

Many researchers have commented on the highly fortuitous correspondence between starlight, more specifically the light of our Sun, and the radiation necessary for photosynthesis. Wald stressed the point in a well-known *Scientific American* article entitled "Life and Light":

> [T]he radiation that is useful in promoting orderly chemical reactions *comprises the great bulk of that of our sun*. The commonly stated limit of human vision—400 to 700 millimicrons [nm] already includes 41 percent of the sun's radiant energy, and about 83 percent of that reaching the earth... The entire photo biological range—300 to 1,100 millimicrons—*includes 75 percent of the sun's radiant energy.*[43]

More recently Thomas Goldsmith commented:

> As L. J. Henderson pointed out... in his book *The Fitness of the Environment* many of the physical features of this planet seem uniquely compatible with life. The mutual dovetailing of biological processes with their physical surroundings is seen nowhere more forcefully than in photobiology.[44]

That the radiation from the Sun (and from most ordinary stars) should be concentrated into a minuscule band of the electromagnetic spectrum that provides precisely the two types of radiation required to maintain life on Earth—light and heat—is surely a very remarkable coincidence. And this is a genuine coincidence, as the compaction of solar radiation into the visible and near infrared is determined by a completely different set of physical laws to those that dictate which wavelengths are suitable for life and photosynthesis.

Others have been far less restrained in admiration about this coincidence than Wald and Goldsmith. The coincidence was described as "staggering" by Ian Campbell in *Energy and the Atmosphere*.[45] As he sums up this remarkable coincidence:

> It just happens that the energy required to excite an electron bound within a molecular orbital to one of the empty orbitals at higher energy is precisely of the order of 10^{-19} to 10^{-18} J [where J = joules]. The important consequence is that the solar radiation impinging upon

a terrestrial molecule can be absorbed in a primary photochemical act and the electromagnetic energy becomes converted into potential chemical energy available to induce secondary reactions. Moreover the typical energy required to break chemical bonds in molecules lies in much the same range... Hence visible light, as it may be termed generally, has exactly the right scale of energy per light quantum or photon to give rise to the possibility of photochemistry, that is chemical reactions driven by the energy of sunlight.[46]

And as he continues:

The Sun just happens to have a surface... temperature of approximately 6000 K, which makes the maximum intensity in terms of photon output within the continuous spectrum lie at around wavelength 600 nm; alternatively the maximum output on an energy basis occurs around 450 nm [i.e., the visual range]... The coincidence [that solar radiation is just right for photochemistry] is the more staggering *when we realize that temperatures well above one million degrees Kelvin must exist within the Sun below the visible surface!*

... What chance then for life to have evolved had the bulk of the Sun's energy output been of quanta in the energy ranges 10^{-20} to 10^{-21} J, or 10^{-18} to 10^{-17} J, or alternatively had the energies required to excite electrons in molecules or break chemical bonds *been in the same ranges?*[47]

Quantum Fitness: Wave-Particle Duality

So far I have in many places talked of electromagnetic radiation in terms of classical wave theory, pointing out that different types of EM radiation, including light and heat, have different "wavelengths" and envisaging the radiation to move through space as ripples move on the surface of a pond.

Many characteristics of light (and other types of EM radiation) do appear to exhibit wavelike behavior, as when light is focused to an image when it passes through a lens. But when light "waves" interact with biomatter, as occurs in photosynthesis when they are absorbed by the chlorophyll molecules in the chloroplast or when they are absorbed by the photo-detector molecules in the retina, the light is detected as a discrete packet or particle of electromagnetic energy known as a photon. In the chlorophyll, the photon raises an electron in the chlorophyll to a higher energy level (see discussion in Chapter 4) and in the photo-detector it changes slightly the structure of a light-sensitive pigment (see further discussion in Chapter 5).

So light behaves at times as a classical wave and at other times as a classical particle. And this strange duality is not restricted to light photons but applies to all subatomic particles, including electrons, protons, neutrons, etc.

In most current texts of biochemistry, the wavelike behavior of matter is seldom alluded to. Atoms and molecules and subatomic particles like electrons and protons are treated almost universally as particles of matter and chemical bonds form and are broken by matter in its particulate manifestation. And this has been the orthodox perspective adopted in biochemistry for the century that has elapsed since the quantum mechanical revolution, at the beginning of the twentieth century, established the so-called "wave-particle duality" of matter. On the whole, biologists have successfully explained most molecular biological phenomena without any appeal to the wave-particle duality of matter.

However, in a cosmos in which there are so many unique and very specific elements of fitness for life on Earth and especially for biological beings of our design, some of which are described in this book (and see also *Nature's Destiny*, *Fire-Maker*, and the *Wonder of Water*), one might have predicted that the weird behavior of matter at the quantum level might also turn out to be uniquely fit in different ways for life on Earth. That this expectation is turning out to be correct is supported by the growing number of recent studies which suggest that quantum phenomena are indeed playing a role in many biological processes.[48]

Wave-Particle Duality. Richard Feynman, one of the most brilliant of twentieth-century physicists, opened the third volume of his classic *Lectures on Physics*[49] with a section headed "Atomic mechanics," containing a succinct description of the weirdness of the behavior of matter at the subatomic level:

"Quantum mechanics" is the description of the behavior of matter and light... on an atomic scale. Things on a very small scale behave like nothing that you have any direct experience about. They do not behave like waves, they do not behave like particles, they do not behave like clouds, or billiard balls, or weights on springs, or like anything that you have ever seen.

Newton thought that light was made up of particles, but then it was discovered that it behaves like a wave. Later, however (in the beginning of the twentieth century), it was found that light did indeed sometimes behave like a particle. Historically, the electron, for example, was thought to behave like a particle, and then it was found that in many respects it behaved like a wave. So it really behaves like neither... There is one lucky break, however—electrons behave just like light. The quantum behavior of atomic objects (electrons, protons, neutrons, photons, and so on) is the same for all, they are all "particle waves," or whatever you want to call them. So what we learn about the properties of electrons (which we shall use for our examples) will apply also to all "particles," including photons of light....

Because atomic behavior is so unlike ordinary experience, it is very difficult to get used to, and it appears peculiar and mysterious to everyone—both to the novice and to the experienced physicist. Even the experts do not understand it the way they would like to, and it is perfectly reasonable that they should not, because all of direct, human experience and of human intuition applies to large objects. We know how large objects will act, but things on a small scale just do not act that way. So we have to learn about them in a sort of abstract or imaginative fashion and not by connection with our direct experience.

In this chapter we shall tackle immediately the basic element of the mysterious behavior in its most strange form. We choose to examine a phenomenon which is impossible, *absolutely* impossible, to explain in any classical way, and which has in it the heart of quantum mechanics. In reality, it contains the *only* mystery. We cannot make the mystery go away by "explaining" how it works. We will just *tell* you how it works. In telling you how it works we will have told you about the basic peculiarities of all quantum mechanics.

Feynman then goes on in the next section, headed "An experiment with bullets," to describe *the double slit experiment* which graphically reveals something of the strange, dual, wave-particle behavior of matter at the subatomic level (see comments below and Figure 2.8).

Feynman again stressed the weirdness of quantum behavior in the introduction to his later book *QED: The Strange Theory of Light and Matter*: "I am describing to you *how* Nature works, you won't understand *why* Nature works that way. But you see, nobody understands that. I can't explain why nature behaves in this peculiar way." And he goes on to say, "I hope you can accept nature as she is—absurd."[50]

The double slit experiment is one of the classic ways in which the dual nature of matter (photons, electrons, or any subatomic parcel of matter) as "particle waves"—as Feynman describes them—can be demonstrated. In this experiment, individual electrons are fired at a piece of apparatus consisting of a double slit positioned in front of a detector screen. The electrons pass through the slits and hit the detector screen behind the slits. As each particle hits the screen, it is detected as a point of light. But although they are detected as points of light individually on the screen, over time they form a set of light bands on the detector screen or what is termed a "diffraction pattern." Their paradoxical dual nature (from the classical point of view or that of our everyday experience) is indicated by the fact that the electrons hit the screen as discrete particles but the overall pattern they generate implies that in passing through the slits they have behaved like a wave (see Figure 2.8).

The possibility that a new era of quantum biology may be dawning is hard to dismiss after watching some of the recent discussions posted on the web.[51] Clearly, interest in the possibility that quantum phenomena play a role in many biological processes is no longer restricted to a tiny fringe minority of biologists.

Quantum Tunneling. One of the mysterious abilities that the wave-like persona of subatomic particles confers on matter at the sub-microscopic level is referred to as "quantum tunneling"[52] (mentioned above in discussing the phenomenon of nuclear fusion). In his paper "Quantum

Tunnelling to the Origin and Evolution of Life," Frank Trixler writes: "The central characteristic of quantum tunnelling is the fact that under certain conditions elementary particles, nucleons or atoms are able to negotiate the obstacle of a potential barrier (which is, from the classical point of view, a forbidden area for a particle) without having the energy to overcome it."[53] The trick is that the "particle" is able to surmount the barrier as a "wave."[54]

Long before the current interest of biologists in quantum phenomena began to take hold, one very important biological consequence of tunneling, that of enabling nuclear fusion to occur in the hot interior of stars, was well-established.[55] As Frank Trixler points out,[56] this in turn enabled the synthesis of carbon and the other higher atoms necessary for life, such as oxygen and iron, and the building of planets. Specifi-

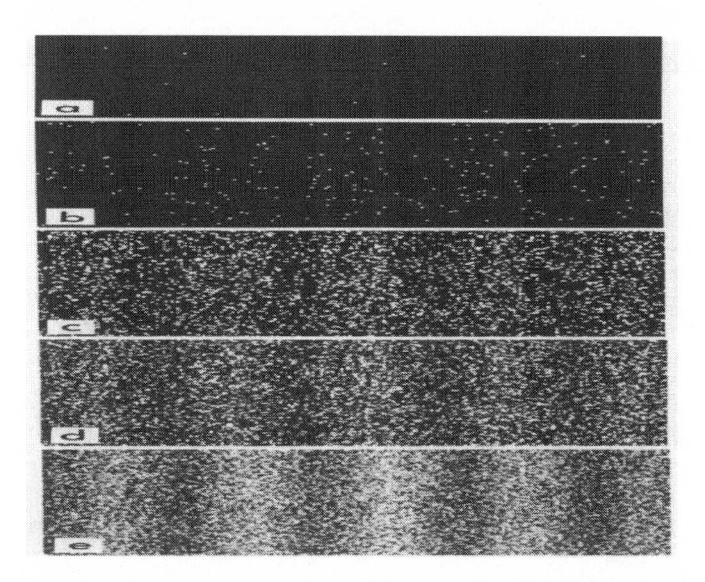

FIGURE 2.8. Results of a double slit experiment performed by Dr. Tonomura showing the build-up of an interference pattern of single electrons. Numbers of electrons are 11 (a), 200 (b), 6000 (c), 40000 (d), 140000 (e). Although the electrons are detected as individual particles when they hit the detector screen (as they are by the chlorophyll in the chloroplast and the retinal pigment in the photoreceptors), their behavior in forming the diffraction pattern of light and dark bands can only be accounted for in terms of "wavelike" behavior.

cally, tunneling allows for atoms in stars to overcome their electrostatic repulsion more often, to combine into greater numbers of new atoms. This process of fusion provides the enormous energies in stellar interiors, which enables stars like the Sun to emit the necessary light and heat for life over billions of years. Remarkably, tunneling is also critically involved in the synthesis of hydrogen (H_2) and water (H_2O) in the deep cold of interstellar space as well as other key organic building blocks of life. As Trixler comments:

> In summary, it can be stated that quantum tunnelling is a key process at the basis of chemical evolution towards prebiotic chemistry. The tunnelling phenomenon does not only play a crucial role for astrochemistry in stellar interiors but also for cold regions of the interstellar medium. In neutral diffuse clouds and especially in dark clouds quantum tunnelling boosts various surface reactions on interstellar dust grains towards the synthesis of important prebiotic molecules.[57]

But in addition to its well-established role in nuclear synthesis, there is growing evidence that tunneling plays a role in many biological processes.[58] For example, when electrons travel down electron transport chains (ETC), such as those which occur in the thylakoid membranes in the chloroplast, they do not hop from carrier protein to carrier protein as discrete particles, but rather tunnel as wave forms along the ETC. As McFadden and Al-Khalili (2013) comment: "Few scientists now doubt that electrons travel along respiratory chains via quantum tunneling."[59] And there is evidence that tunneling is also occurring in many other protein functions and enzymic reactions.[60] And it is particularly fascinating that this mysterious process is also implicated in photosynthesis—more specifically in the transfer of excited electrons through the forest of chlorophyll molecules in the thylakoid membrane.[61]

As McFadden and Al-Khalili explain in their book *Life on the Edge*, if a plant is to harness its captured solar energy efficiently, it has to ensure that the excited electrons transit very rapidly through the chlorophyll forest by taking the shortest route to the reaction centers.[62] As the authors relate, recent research suggests that the electron may find its way through the chlorophyll forest by adopting a wave form and tunneling through. It is surely an intriguing possibility that the quantum properties of matter and its ability to "tunnel" at the subatomic level are of utility (perhaps crucial) in what is, for us oxygen-hungry aerobes, the most important chemical reaction on planet Earth. And it is fitting that the wave-like properties of light are fit for an entirely different phenomenon, but also of profound importance to ourselves, the formation of a high-acuity image on the human retina. (See discussion in Chapter 5.)

Quantum tunneling may seem "impossible" or "absurd" as Feynman claimed, but it seems that it is fit for life in a vast number of diverse processes. Frank Trixler captured something of its ubiquitous fitness for life in the conclusion of his paper:

> This review was a round trip from hot stellar interiors to cold interstellar medium, from deep lithospheres and subsurface oceans to planetary upper atmospheres and from the microcosm of biomolecular nanomachines to the evolution of macroscopic multicellular life. At each stop we have seen that quantum tunnelling is of vital importance for life and its origin. These stops are not isolated but rather closely connected with each other. The following example further clarifies this concept by focusing on photosynthesis as an example.
>
> Photosynthesis is basically performed by functional biomolecules whose redox reactions rely on long-range electron tunnelling. All elements which build these biomolecules (except hydrogen) were produced by thermonuclear reactions in stars via nuclear tunnelling... the energy source of oxygenic photosynthesis is sunlight which is a product of nuclear tunnelling in thermonuclear reactions. Our sun itself was formed by the collapse of dark interstellar clouds. In this process, molecular hydrogen plays an important role and is formed in dark clouds by the help of quantum tunnelling. Oxygen, the important byproduct of oxygenic photosynthesis, is used by multicellular lifeforms for cellular respiration—a process which relies on long-range electron tunnelling. The evolution of all multicellular lifeforms is based on DNA mutations and energy input over bil-

> lions of years. A source of mutations is proton tunnelling in DNA while the large period of time of solar energy input is due to specific characteristics of quantum tunnelling taking place in thermonuclear reactions.
>
> Although this example does not cover every single topic previously discussed, it illustrates the main message of this review: *there is a highly multidisciplinary network of quantum tunnels essential for the origin and evolution of life.*[63]

The mysterious ability of particles to adopt the form of a wave turns out to be fit for life in a number of very diverse ways. Without quantum tunneling there would be no nuclear fusion, stars would never heat up, and no life-giving light or heat would flood the cosmos and engender life on the surface of a planet like the Earth. There would be no carbon nor any other higher atoms in the universe, and certainly no photosynthesis nor any oxygen to satisfy the metabolic needs of advanced organisms like ourselves.

Feynman lamented that no one knows "why nature behaves in this peculiar way." But I think it is increasingly clear that *without* nature behaving in "this peculiar way," manifest in the ability of subatomic matter to exhibit both particle- and wave-like behavior, the cosmos would be a very different place. If Nature did not behave in "this peculiar way" the cosmos would not only be a very different place, it would be devoid of carbon-based life and certainly devoid of any advanced carbon-based life forms like ourselves with the gift of sight (see Chapter 5).

Fitness of EM Radiation Outside the Visible Region

As we saw in the first part of this chapter, the only electromagnetic radiation of any utility to biological systems is restricted to the tiny visual and IR bands. However, many regions of the EM spectrum outside these two bands have proven to have many useful technological applications. Similarly, many atoms of the periodic table that have, as far as we know, no utility for biological systems, have many important, intriguing, and sometimes very specific industrial and technological applications, such as uranium (atomic power), lithium (batteries), neodymium (powerful magnets in electric motors), and many of the rare Earth metals.[64]

As pointed out in *Encyclopaedia Britannica*, microwaves are used in every mobile phone and in Wi-Fi devices.[65] Similarly, radio frequencies have been used for communication for more than 100 years, as through wireless and TV. X-ray crystallography[66] was used to discover the structure of proteins and nucleic acids. It was the X-ray crystallography carried out by Rosalind Franklin at King's College that played a decisive role in the discovery of the double helix.[67]

I have a personal affection for the role played by researchers at King's College, London, as I was a student there myself shortly after the X-ray diffraction work that led to the elucidation of the structure of the double helix. I was privileged to use the same specially built vibration-free cold room, in the basement at King's, used to form the crystals used by Rosalind in her X-ray diffraction work, for my own PhD studies on the development of the red cell.

Other technological uses of EM spectra: X-rays are used extensively in medicine, in simple X-rays and in CAT scans. Nuclear Magnetic Resonance (NMR), another important medical technique, uses a radio frequency pulse. And many scientific techniques outside of medicine exploit NMR to study molecular physics, crystals, and non-crystalline materials. Again UV lasers have various industrial applications.[68]

And of course, X-rays, UV radiation, microwaves, and radio waves have been used along with light astronomy to vastly increase our understanding of cosmic and stellar evolution and of the overall structure of the cosmos.

The utility of so many regions of the electromagnetic spectrum outside the tiny visual and IR bands, in conjunction with the technological utility of so many atoms which as far as we know serve no functions in biological systems, lends further credibility to the argument I put forward in *Fire-Maker*, that nature is not just fit for our biological being but also seemingly prepared for intelligent beings of our biology to develop an advanced technological civilization.

Conclusion

Because we and all other complex, advanced life on Earth derive energy from the reaction of oxygen with reduced carbon compounds which are themselves (the O_2 and CH) the product of photosynthesis, we are not just "star-stuff," as Carl Sagan pointed out[69] (referring to the fact that the atoms of our body, including oxygen and carbon, are synthesized in stellar interiors), but also children of the light, or "light eaters" as described so evocatively by William Broad.[70]

To us "light eating" aerobic chauvinists, living on the proceeds of photosynthesis and having been taught at school the mantra that *all life depends on the Sun,* it comes as something of a shock to learn that our aerobic, photosynthesis-dependent, extravagant lifestyle is followed by only a small fraction of all living species and perhaps less than half of the biomass on Earth—much of which occupies subsurface ecological niches penetrating kilometers into the lithosphere (see Appendix A, "Doing Without Sunlight"). We in fact belong to a unique category of light-dependent forms of carbon-based life, one among a multitude of diverse energy-harvesting lifestyles, deviants in a far greater, more inclusive kingdom of life, many of whose denizens are anaerobic and have no need for the nutrients gifted by photosynthesis.

And this implies something more. The astonishing compaction of the radiation emitted by our Sun (and most stars in the universe) in the tiny visual and infrared bands, and the coincidence that this exceedingly narrow portion of the electromagnetic spectrum should be precisely the radiation required for photosynthesis (to excite molecules for chemical reaction; not too energetic, not too weak, just right[71]), for providing the necessary heat to warm the atmosphere, for preserving water as a liquid on the Earth's surface, and for animating molecules for chemical reactions, *is* a fitness *specifically for us "light eaters"*—the advanced carbon-based life forms that inhabit the surface of the Earth. *It is for us that this coincidence is of such crucial importance.*

The fact that the tiny band of visual light and the tiny band of heat we need for life on Earth coincide with the tiny band emitted by our Sun is surely a fact of supreme importance and lends credibility to the idea that the universe is particularly friendly for life like us.

But it is not enough for the Sun to emit the right kind of light. The light must reach the Earth in order to be useful to us. That is what we will explore in the next chapter.

FIGURE 3.1. Sunlight reaching through the atmosphere.

3. Letting the Light In

> [L]et us imagine that intelligent life once evolved on [a cloud-covered planet like Venus]. Would it then invent science? The development of science on Earth was spurred fundamentally by observations of the regularities of the stars and planets. But Venus is completely cloud-covered... nothing of the astronomical universe would be visible if you looked up into the night sky of Venus. Even the Sun would be invisible in the daytime; its light would be scattered and diffused over the whole sky—just as scuba divers see only a uniform enveloping radiance beneath the sea.
>
> —Carl Sagan (1980), *Cosmos*[1]

It is remarkable enough that the Sun and the vast majority of stars beam out EM radiation mainly in the tiny Goldilocks region of the EMS. But to enable life on the Earth's surface—or indeed on any rocky planet anywhere in the universe—a further precondition must be satisfied. The life-giving light of the Sun must penetrate the atmosphere right down to the ground to work its magic, and a proportion of the Sun's IR radiation (heat radiation) must be absorbed by and held in the atmosphere to warm the Earth above the freezing point of water and animate the atoms of life for chemistry.

Amazingly, the atmosphere obliges us in this critical task. But as we shall see, its capacity to let through the right light and absorb the right proportion of heat depends on an additional suite of hugely improbable coincidences in the combined absorption characteristics of the atmospheric gases.

Letting the Light Through

In the first stunning coincidence, upon which our type of "light eating" lifestyle is critically dependent, our atmosphere obligingly lets through to the Earth's surface most of the right light for photosynthesis (the prime biological process upon which we "light eaters" depend)—that is, light in the visual region of the EMS, between wavelengths of approximately 0.3 microns and 0.8 microns. The atmosphere also lets through a significant fraction of radiation in the near IR region to the Earth's surface (felt as warmth on the skin) while at the same time retaining a fraction in the atmosphere, raising the temperature of the Earth into the ambient temperature range, preserving liquid water on the Earth's surface as well as (see below) animating molecules and atoms for chemical reactions.

Overall, the absorbance properties of the Earth's atmosphere with regard to electromagnetic radiation in the visual and IR regions are just right for photosynthesis and just right to raise the temperature of the atmosphere of our planet into the ambient temperature range enabling "light eating" aerobes like ourselves to thrive on the surface the Earth.

It is not just that the atmosphere lets through the right light. It also strongly absorbs radiation from the dangerous or potentially dangerous regions of the EM spectrum on either side of the visual and near IR regions (see Figures 3.2 and 3.3). The only other EM radiation that is not

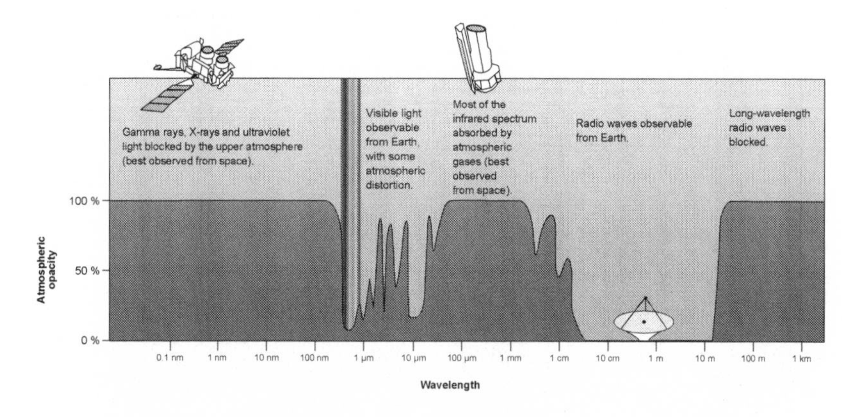

FIGURE 3.2. Absorbance of EM radiation by the atmosphere. X-ray region from 0.1-10 nm; UV from 10 nm–400 nm; visual region about 350 nm–800 microns [some animals can see UV so the visual range extends into the UV region]; infrared region from 800 nm–1 mm ; microwave 1mm to 1 meter; radio region beyond 1 meter..

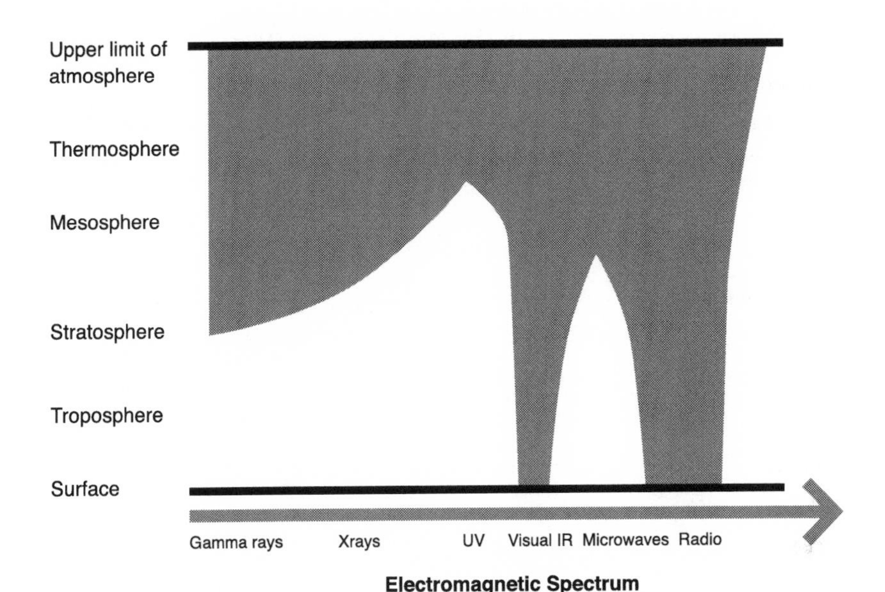

FIGURE 3.3. Penetration of EM radiation into the atmosphere before absorption. Only radiation in the near UV, visual and near IR (the optical window) and in the far microwave and a portion of the radio region (the radio window) reaches the ground. Although some microwave radiation in the far end of the band (beyond five centimeters) does penetrate to the ground, the amount of microwave region emitted by the Sun in the far microwave region is absolutely minuscule. Intriguingly, while X-rays and UV have many technological applications that do not necessitate atmospheric transparency, many of the applications of microwave and radio waves are critically dependent on the existence of the radio window.

absorbed is in the radio region and the far microwave region (see Figures 3.2 and 3.3). Consequently, virtually no strong or ionizing radiation in the UV, X-ray, and gamma-ray regions—that is, radiation less than 0.3 microns—penetrates to the Earth's surface. Further, absorbance by the atmospheric gases of radiant energy of wavelengths longer than fifteen microns in the far IR and near (or high end) microwave regions means that virtually no radiation in these regions of the EM spectrum reaches the ground (see Figures 3.2 and 3.3). This is also almost certainly protective because, as discussed in note 40 of the last chapter, microwave radiation has many reported damaging effects on living systems even at very low radiation fluxes.

Just how lucky we are that the useful radiation gets through is highlighted by the strong absorption in the bands immediately adjacent to the visual and near infrared bands; in the far UV and in the far infrared and microwave (cooking wavelengths)—regions of the spectrum *just slightly outside the right region*!

As can be easily envisaged, if the absorbance by the atmosphere had covered a slightly different region of the EM spectrum, if we imagine it shifted from its current position by even a slight degree—to the "right" for example (see Figure 3.2), so that the atmosphere absorbed all the visual light and all the IR and let through instead the adjacent far UV (which is lethal to life), then not only would photosynthesis have been impossible, but the world would have suffered a runaway greenhouse effect and ended up a hot hell-house like Venus because of the complete absorption of all the Sun's IR radiance. In such a scenario, no carbon-based life could have survived on the surface of the Earth and certainly no air-breathing aerobes like ourselves! Or conversely, if we imagine it shifted to the left, all the light and IR would have been absorbed by the atmosphere, again causing a runaway greenhouse effect.

Again, as in the case of depictions in the literature of the regions occupied by the visual and IR regions in the EM spectrum (discussed in the previous chapter), depictions in the literature showing atmospheric transmission of EM radiation restricted to a narrow band comprising the near UV, visual, and near IR regions—and Figure 3.2 and 3.3 are typical in this regard[2]—fail to capture the fact that the radiation in these regions represents again a tiny, unimaginably small fraction of the entire span of the EM spectrum. A more accurate depiction is shown in Figure 3.4.

That the EM radiation emitted by the Sun *and* the EM radiation allowed through the atmosphere should both be largely restricted to the *same* tiny useful regions—providing light for photosynthesis (to activate matter for chemical reaction) and heat to warm the earth and bring

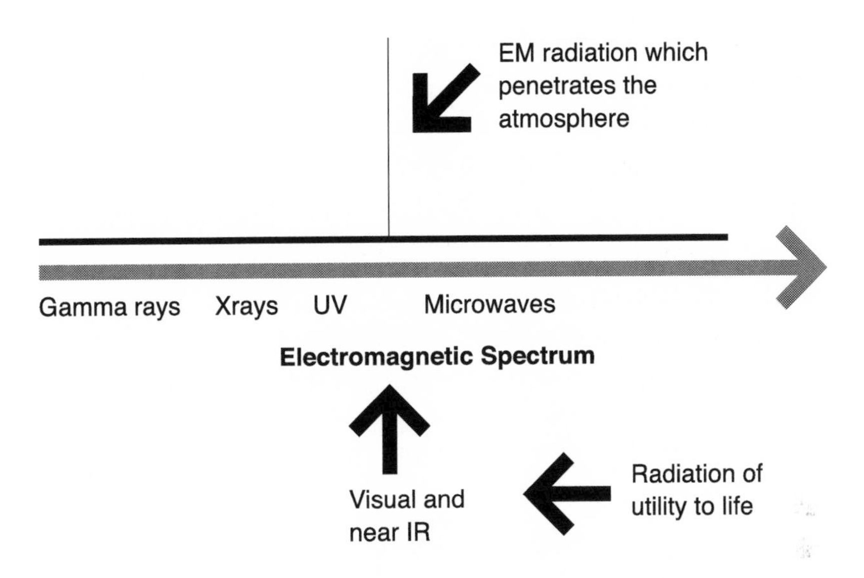

FIGURE 3.4. EM radiation penetrating the atmosphere.

molecules into contact (as discussed in the previous chapter) so that chemical reactions can occur—is an extraordinary example of a special fitness in nature for our type of aerobic life on a planetary surface. The correspondence is simply stunning. If I can be excused for expressing the coincidence in animist terms, it is as if the atmosphere were intelligently colluding with the Sun to ensure that only the right light for photochemistry—between wavelengths 0.3 microns and 0.8 microns—reached the Earth's surface and that only the 'right' proportion of the IR was absorbed to warm the Earth into the ambient temperature range.

Water's Transparency to Visual Light. Despite the vital life-giving transparency of the atmosphere to light in the visual region of the spectrum, the life-giving light of the Sun would not be of utility for photosynthesis without another unique element of fitness—the absorption spectrum of liquid water. As shown in Figure 3.5, water is nearly completely transparent to visual light, while nearly all electromagnetic radiation adjacent to the visual band, including the far IR and UV,[3] is strongly absorbed. Some near IR does penetrate water,[4] as is witnessed by the fact that the warmth of direct sunlight can be felt on the skin in the upper

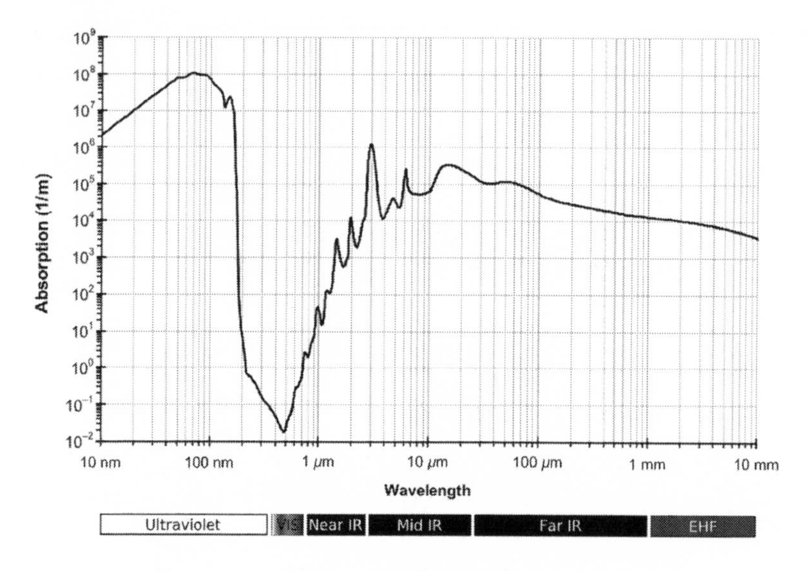

FIGURE 3.5. Absorption of EM radiation by water.

few centimeters of a body of water. Again, like the atmosphere, water not only lets the right light through, but shields life from all the dangerous EM radiation—in the far UV and X-ray regions as well as in the far IR and potentially dangerous microwave regions.

Moreover, water is transparent to visual light not only as a *liquid,*[5] but also as *vapor* in the atmosphere[6] and as *ice.*[7] If liquid water or water vapor in the atmosphere absorbed visual light—the right light for photosynthesis—*then photosynthesis would not be possible, and Earth would be devoid of aerobic life forms.*

Yet again, depictions in the literature of the absorption spectrum of liquid water such as that shown in the *Encyclopedia Britannica*[8] and in Figure 3.3, give the values of the wavelengths in logarithms, and this fails to highlight the exceeding narrowness of the window in the absorption spectrum of water which lets through just the right light. In reality, the window is an almost infinitely thin line in the immensity of the EM spectrum.

The significance of the transparency of water to visual light can hardly be exaggerated. All biological chemistry occurs in liquid water. If the energy of sunlight is to sustain photosynthesis in the ocean or in rivers and lakes, it must be capable of penetrating some distance below the surface of the water. Even on land, if light energy is to reach the chemical machinery in the chloroplasts of terrestrial plants, the precious photons must invariably penetrate a thin layer of water.

Absorbing the Heat

THE ABSORPTION of IR by the atmosphere differs critically from the absorption of radiation in the visual region. Although the atmosphere lets through much of the Sun's incoming IR radiation (in the 0.8–5 micron region) to the Earth's surface through a series of absorption windows (see Figure 3.2)—which as mentioned above is felt as warmth on the skin—it also absorbs, via a series of absorption bands, a proportion of the incoming IR radiation and much of the re-emitted longer IR radiation in the 5–100 micron region.[9]

IR absorption bands. It is hard to emphasize the enormous importance of the very different way in which the atmosphere handles light compared to IR radiation, witnessed in the contrast between the strong absorption bands in the IR compared with their absence in the visual region. All aerobic life inhabiting the Earth's surface hangs on this crucial difference. The fitness of nature for photosynthesis depends critically as much on some of the IR radiation being absorbed as it does on the visual radiation *not* being absorbed.

If the same non-absorbance and maximal penetration were also true of the radiation in the adjacent near IR (as it is in the visual region), the results would be quite calamitous. Without some absorbance by the atmosphere when the Sun was shining, it would be intolerably hot, while as soon as night fell, the temperature would suddenly fall below zero. We would experience diurnal temperature variations like those on the Moon. Temperatures on the Moon are very hot in the daytime, about 100°C. However, at night, the lunar surface gets very cold and the temperature drops to –178°C. This wide variation occurs because the Moon has no atmosphere—no greenhouse blanket—to retain heat at night or prevent the surface from getting so hot during the day.[10] Without the greenhouse blanket, even walking from an area in direct sunlight to a shaded area would be like walking from a hothouse into a fridge. No type of carbon-based plant life instantiated in a water matrix could survive in such a hostile, alien environment, with such massive temperature fluctuations. It is the absorption of IR radiation by the atmosphere which saves us and all life on Earth from such fluctuations.[11] During the day, the absorption protects us from the Sun's heat, while at night, the heat retained in the atmosphere prevents a calamitous fall to temperatures way below zero.

Overall, the absorption and retention of the Sun's heat in the atmosphere raises the mean global average temperature by 33°C above what it would be without the greenhouse blanket, from a chilly –18°C to 15°C.[12]

But *too much* absorption in the IR region would be equally calamitous, leading to a runaway greenhouse. The Earth is saved from this fate because of the absorption windows in the 1–15 micron region. These are as crucial for life on Earth as the absorption peaks. Why? Because without some windows, all the incident IR radiation would be absorbed by the atmosphere and none could be radiated back out into space. If that were the case, the Earth would suffer a runaway greenhouse effect and end up like Venus. What is needed in the IR region is exactly what we see: A number of windows, interspersed with a number of absorption peaks, so that in the end only a fraction of the incident IR reaching Earth is absorbed in the atmosphere and only a fraction of the IR re-emitted is absorbed.

A very specific and intriguing detail in the absorption characteristics of the atmosphere in the IR region is the existence of a large absorbance gap or window between eight and fourteen microns.[13] Why is this intriguing? Because the Sun is not the only body that emits IR radiation. The Earth also emits some IR radiation, since all bodies at a given temperature emit radiation with a characteristic range of wavelengths. In the case of the Earth, the peak of emission is in the IR region.[14]

The absorption gap allows a considerable proportion of the Earth's IR emission to escape into space through the large eight-to-fourteen-micron window. Roughly estimating from the figure shown of the Sun's radiation spectrum, given in John Mitchell's paper "The 'Greenhouse' Effect and Climate Change,"[15] almost twenty-five to thirty percent of the long IR emission from the Earth escapes through this window. Clearly this particular window is playing a major role in preventing a runaway greenhouse effect on Earth. It may not be an exaggeration to claim that despite all the fortuity in nature which makes possible our existence as children of the light, without this window—just one tiny detail in the overall absorption spectrum of the atmosphere—we would not be here. And if so, this would represent a staggering example of the bio-centric fine-tuning of nature.

But aside from this, what is certain is that if *all* radiation in the infrared between 0.80 and 100 microns had been absorbed by the atmospheric gases, if there were no windows, a runaway greenhouse would have been inevitable. The Earth would be a hot, Venus-like planet. On those windows, including the eight-to-fourteen-micron window, all advanced life on the surface of the Earth depends.

The Composition of the Atmosphere

The absorbance characteristics of the Earth's atmosphere are the combined result of the absorbance spectra of five gases, four of which—nitrogen (N_2), oxygen (O_2), carbon dioxide (CO_2), and water vapor (H_2O)—make up the bulk of the atmosphere, and the fifth of which—ozone (O_3)—is present in only trace amounts.

The fact that the combined absorbance characteristics of these five gases provide just the right absorbance characteristics necessary for advanced aerobic life on the earth's surface, letting through the right light for photosynthesis and absorbing sufficient heat to raise the earth's temperature to within the ambient range, is an extraordinary fact—one of the most astonishing elements of fitness for life in all nature. Why? Because the five atmospheric gases N_2, O_2, O_3, H_2O, and CO_2, four of which —N_2, O_2, H_2O, and CO_2—form the bulk of the atmosphere, must exist on any planet hosting complex carbon-based biological life. That their absorbance characteristics should be of such vital benefit for life is therefore a coincidence of stunning fortuity.

Oxygen. Oxygen (O_2) is essential for complex organisms like ourselves. We need it in copious quantities (250 milliliters *every minute,* even at rest).[16] And advanced carbon-based life anywhere in the universe is bound to use oxidations to generate energy and take it up directly from an atmosphere.[17] (That the highest metabolic rates needed to sustain advanced complex life depend on being air-breathing, that is, taking oxygen directly from an atmosphere, is discussed in Chapter 4.) Consequently, *atmospheres sustaining complex life will inevitably contain a considerable proportion of oxygen.*

Ozone. And where there is oxygen there is bound to be ozone (O_3), which is formed in trace amounts in the stratosphere by the action of UV light on molecular oxygen.[18]

$$O_2 + O = O_3$$

Carbon Dioxide. As you sit reading this text, you are not just taking in oxygen *from* the atmosphere; you are at the same time breathing out carbon dioxide (CO_2) *into* the atmosphere. CO_2 is one of the major products of cellular respiration, the process which provides us with ninety percent of our energy needs. Moreover, CO_2 is not only being delivered continuously to the atmosphere by all aerobic organisms as the inevitable waste product of respiration, but is also, as Henderson pointed out, the *only feasible carrier molecule of the carbon atom* to all parts of the biosphere, because it is a gas that is readily soluble in water.[19] So CO_2 is bound to be a component of the atmosphere of any world in which carbon-based life is established and where aerobes are utilizing the oxidation of reduced carbon for the generation of metabolic energy. In addition, CO_2 is continually being delivered to the atmosphere on Earth by volcanic activity and recycled by silicate weathering (see below).

In passing, it is worth noting that the dual role of CO_2, both as the carrier of the carbon atom to every corner of the biosphere (including the leaves of terrestrial plants) and as an absorber of IR radiation in the atmosphere, is only possible because CO_2 *is a gas at ambient temperatures*. This is another fortuitous and critical element of fitness in nature for carbon-based life on Earth because, as Arthur Needham pointed out: "It is in fact one of the very few gaseous oxides at ordinary temperatures."[20]

So O_2, O_3, and CO_2 are necessary constituents of the atmosphere of any world harboring advanced carbon-based life.

Water Vapor. The presence of water vapor (H_2O) in the atmosphere of worlds harboring carbon-based life is also necessary for reasons besides its absorption characteristics. Water is the essential physical matrix of the carbon-based cell and is present in vast quantities on the surface of the Earth in lakes, rivers, and the oceans. Water evaporates at ambient temperatures (the mean ambient temperature of Earth is 15°C, as mentioned above) and is bound to be a constituent of the atmosphere of any world bearing our type of life.

Nitrogen. Atmospheric nitrogen is the major source of all the nitrogen atoms incorporated into the organic compounds, including DNA and proteins, of living organisms on Earth. After plants have reduced the carbon in CO_2 to organic compounds containing C, H, and O, the next step in organic syntheses is the incorporation of nitrogen to form the vast inventory of organics with the generic formula CHON, which includes proteins and nucleic acids.

As well as being one of the core four atoms of organic chemistry (C, H, O, and N), nitrogen provides density to the atmosphere, preventing the oceans from evaporating. Nitrogen also acts as a crucial diluent, quenching fire (see *Fire-Maker*) and slowing the speed at which fire spreads, rendering it controllable at the O_2 levels necessary to support human respiration (150 mm Hg). It is again the only available candidate for this role and is likely therefore to be an essential component in the atmosphere of all planets harboring advanced aerobic life.

All this implies that the aforementioned atmospheric gases—oxygen, nitrogen, water vapor, carbon dioxide, and ozone—are inevitable on any world like our own, inhabited by photosynthesis-dependent, oxygen-utilizing, advanced carbon-based life, *for reasons of biological necessity which have nothing to do with their collective absorbance properties!*

Moreover, even the relative proportions of these gases must be approximately what they are to sustain a biosphere bearing advanced life. In the case of oxygen, only a concentration of about twenty percent at a partial pressure of 150 mm Hg will provide sufficient oxygen to support the active metabolism of organisms like ourselves. On the other hand, if the oxygen content of the atmosphere were higher, then fire would be a far more dangerous phenomenon. In the case of nitrogen, only a considerable quantity of nitrogen will provide the density and pressure necessary to keep the oceans from evaporating. Considerable amounts of nitrogen are also necessary to act as a fire-retardant, and in an atmosphere containing 20 percent oxygen a level of about 80 percent is necessary for quenching fire.[21] (See the discussion in Chapter 2 of *Fire-Maker.*[22])

In the case of water vapor, given the ambient temperature range on Earth and the vast area of the oceans and lakes, there is bound to be some water vapor in the atmosphere, although the amount will vary in different places depending on temperature and geography (the amount of water vapor over tropical bodies of water will be more than the amount over a desert). Similarly, CO_2 levels have varied throughout geological time, although over the past 400 million years—since advanced life emerged from the sea—they have probably never reached levels ten times present levels and probably never more than about four to five times present levels.[23] Some intriguing circumstantial evidence for this was shown in a recent study, that in the case of humans, raising CO_2 levels in controlled atmospheres up to four times current levels causes a diminution of cognitive function.[24] This suggests there may be a ceiling on the level of CO_2 compatible with advanced life.

Because of the vast amounts of oxygen in the atmosphere there is inevitably going to be a small amount of ozone. Ozone is a very powerful greenhouse gas that absorbs strongly in the IR region—1,000 times more powerful than CO_2.[25] Because of this, any more than trace amounts would contribute dangerously to the greenhouse effect. Its life-giving fitness in absorbing the dangerous UV between 0.20 and 0.30 microns[26] would be negated entirely if more than trace amounts were necessary. Significantly, the rate of breakdown of ozone in the stratosphere almost equals its rate of synthesis, guaranteeing that it is indeed only present in trace amounts.

To repeat: the presence of these gases in the atmosphere—and their relative proportions—are the result of processes and phenomena which are totally unrelated to their magic collective absorbance properties reviewed above. Their specific absorption properties are therefore an element of fitness in nature of immense fortuity, a coincidence of vital significance which makes possible photosynthesis (letting through the light) and keeps the temperature of the Earth within the ambient temperature range (absorbing a fraction of the IR).

The Greenhouse Gases. A critical characteristic of atmospheric fitness regarding the absorption of IR radiation is that diatomic molecules consisting of the same two atoms, like O_2 and N_2, do not absorb IR radiation. This is another highly fortunate element of natural fitness for life on Earth. If either of the two gases which make up the bulk of the atmosphere, O_2 and N_2, had been strong absorbers of infrared radiation, the Earth might have become a hothouse like Venus, where the temperature on the surface of the planet is hot enough to melt lead.[27]

The fact that oxygen is *not* a greenhouse gas is important for another reason. The only major change in the composition of the atmosphere over the past three billion years was an increase in oxygen levels, from vanishingly small amounts two billion years ago to the present level, which was reached several hundred million years ago (see Figure 3.6). Since then, it has fluctuated between about fifteen and thirty-five percent.[28] Intriguingly, this change—the only major long-term change in the composition of the atmosphere in three billion years—is critically related to our own existence, as it was only this crucial change that permitted the use of oxygen as a source of energy and the subsequent advent of advanced organisms with high metabolic rates like ourselves.

As mentioned above, oxygen, which is vital for all advanced life forms on Earth, makes up about twenty-one percent of the current atmosphere (pO_2 of 150 mm Hg). Levels close to this are necessary to support oxidative metabolism in metabolically active organisms. If oxygen had been a greenhouse gas like water vapor or carbon dioxide, the result would have been catastrophic. Photosynthesis might have started, but soon the buildup of oxygen would have cooked the Earth and complex life would have been obliterated. Oxygen, being a diatomic gas and *not a greenhouse gas,* is therefore a major element of fitness in nature for advanced life. And while oxygen does give rise to a greenhouse gas, ozone, it does so in only vanishingly small quantities in the upper atmosphere, sufficient to absorb incoming UV but not enough to alter the heat balance of the atmosphere.

Another important property of the greenhouse gases H_2O and CO_2 is that both are stable in the presence of O_2. This is a fact of huge significance. If any of these two gases were unstable in the presence of oxygen, the whole atmospheric system would break down. Aerobic life, our sort of life, would be impossible. However, in keeping with nature's profound fitness for advanced life as it exists on Earth, H_2O and CO_2 are fully oxidized and stable in the presence of oxygen. Nitrogen, the major component of the atmosphere, is also stable in the presence of oxygen because the nitrogen atoms in N_2 bond with each other very strongly and resist combining with oxygen. The unreactivity of oxygen with water, carbon dioxide, and nitrogen is a point worth emphasizing, as most other atoms and organic molecules (apart from the noble gases) react strongly, often explosively, with oxygen.

In sum: The crucial heat necessary to keep the temperature of the Earth in the ambient temperature range, conserve liquid water on Earth, and animate matter for chemical reactions, including those of photosynthesis, is absorbed in the atmosphere by CO_2 and H_2O, and it is only because O_2 and N_2 *do not absorb* IR that the Earth avoids a runaway greenhouse.

Note again: The fact that these two gases make up the bulk of the atmosphere is of necessity because of their vital roles in biological processes—respiration in the case of oxygen, quenching fire (among other roles) in the case of nitrogen; yet these roles, which necessitate their presence in the proportions that they currently occupy in the earth's atmosphere, have *nothing to do with their fortuitous transparency to IR.*

Silicate Weathering.[29] If it can be believed, H_2O and CO_2 are not just responsible for *retaining heat* in the atmosphere. They are also responsible for regulating the *amount of heat* retained in the atmosphere over geological time spans by a feedback process—silicate weathering—another process as far removed from their collective absorption properties as can be imagined!

Very briefly (see Chapter 3 in *The Wonder of Water* for a more detailed description): If the CO_2 level *or* temperature falls, the rate of silicate weathering decreases, increasing the amount of CO_2 in the atmo-

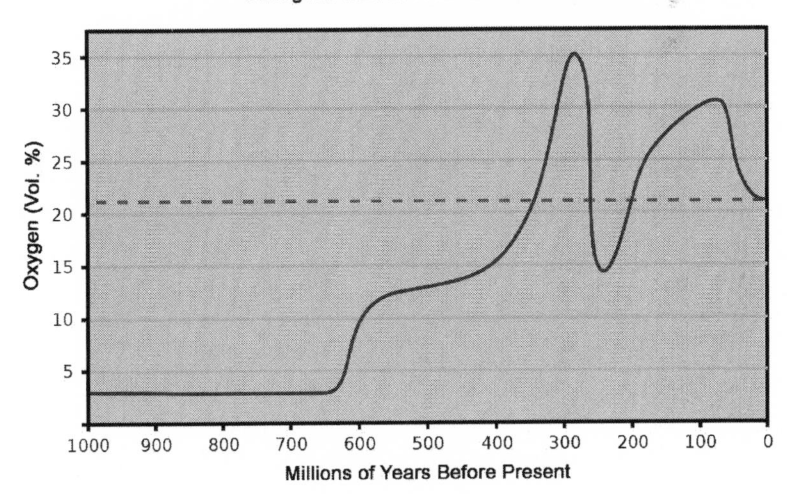

FIGURE 3.6. The oxygen content of the Earth over the last billion years.

sphere and ultimately raising temperatures by the greenhouse effect. If the temperature or the CO_2 level rises, the weathering reaction rate increases, lessening CO_2 levels and the greenhouse effect and lowering global temperatures. This intriguing *negative feedback* mechanism acts to lower or raise global temperatures and CO_2 levels in the atmosphere, returning the Earth to its long-term norm. By maintaining global temperatures near the current ambient range for billions of years, it has also played a vital role in the preservation of the oceans.

This critical negative feedback system involves the chemical weathering of silicate rocks by carbonic acid. As shown in the formula below, as CO_2 levels rise the reaction is driven strongly to the right.

$$Mg_2SiO_4 + 4\,CO_2 + 4\,H_2O \Rightarrow 2\,Mg^{2+} + 4\,HCO_3 + H_4SiO_4$$

[Olivine + carbon dioxide + water ⇨ magnesium and bicarbonate ions in solution + silicic acid in solution.]

The bicarbonate generated (from atmospheric CO_2) is then carried to the oceans, where marine microbes utilize it to manufacture the carbonate that composes their shells, which rain to the sea bed as the organisms die. There it is sequestered in marine sediments carrying the carbon dioxide from the atmosphere into the Earth as the oceanic crust subducts into the mantle. (Note: the activities of living things are therefore also playing a critical role in this remarkable regulatory system.[30])

The net effect is the removal of CO_2 from the atmosphere and the lowering of global temperatures.

On the other hand, if temperatures fall or if the CO_2 level decreases for some reason, then the rate of the reaction is decreased and rate of removal of CO_2 is decreased.

$$Mg_2SiO_4 + 4CO_2 + 4H_2O \Rightarrow 2Mg^{2+} + 4HCO_3 + H_4SiO_4$$

Thus, CO_2 levels increase and eventually raise the global temperature.

Are there two processes in nature as different as silicate weathering and atmospheric absorption of EM radiation? Yet it is only because of the profound fitness of these two major reactants in photosynthesis (CO_2 and H_2O) in enabling *both* these very different processes—heat retention (via atmospheric absorption) and heat regulation (via silicate weathering)—that there has been liquid water on Earth, and photosynthesis has proceeded on Earth over 3.5 billion years!

To repeat: Carbon dioxide and water, both so crucial in other ways for biological life, not only absorb the necessary heat to animate reactions and keep the water on the Earth's surface from freezing solid, but also *regulate* the Earth's temperature, keeping it within the Goldilocks ambient temperature range, not too cold, not too hot, adding to the staggering suite of elements of fitness in nature for life on the surface of the Earth.

The Vital Coincidences

The evidence discussed in the previous chapter and this chapter reveals that life on Earth, and more specifically advanced life living off the proceeds of photosynthesis, depends on a genuine and extraordinary series of coincidences in nature's order.

First, that the two types of electromagnetic radiation of utility to life—light and IR—occupy two exceedingly tiny regions in the immensity of the electromagnetic spectrum, which happen to be the same regions in which the Sun and indeed the vast majority of stars emit nearly all of their radiation.

Second, that the gases of the atmosphere which are necessary for biological reasons let through most of the useful light radiation emitted by the sun to the Earth's surface, where it enables photosynthesis, while absorbing a fraction of the IR radiation to raise the temperature of the Earth's surface into the ambient temperature range.

The *first* is a genuine coincidence, because the laws of nature that determine that light photons have just the right energy levels for photochemistry and that radiation in the IR region has just the right properties to warm the Earth *have nothing to do* with the laws which determine that stars of surface temperature close to 6,000°C (the majority) will emit EM radiation in these *same* two vital bands.

The *second* is again a genuine coincidence, because the laws of nature which determine the spectral absorbance of the atmospheric gases have *no connection with* the biological laws of nature which determine which gases will make up the atmosphere and in which proportions they will occur.

To repeat: The laws of nature which determine the *physical absorbance properties* of the gases have no connection with their *chemical properties* which determine their utility to life.

In the case of photosynthesis, three of the key atmospheric gases whose physical absorption properties are of such critical importance (letting through the light) in enabling the process of photosynthesis to proceed are *also chemical players in the process of photosynthesis itself.*

$$6\,CO_2 + 6\,H_2O + \text{light} + \text{heat} \Rightarrow C_6H_{12}O_6 + 6\,O_2$$

And these gases aren't just peripheral players in the process—they are *the major reactants*! In other words, it is the three major reactants, CO_2, H_2O, and O_2, which ensure—by their collective absorption properties in the atmosphere—the availability of the vital light energy necessary to drive the reaction to completion. It is as if these three gases were colluding intelligently together to promote their incorporation into the substance of living matter.

Altogether these coincidences convey an overwhelming impression of design. The improbability that they are the outcome of the blind concourse of atoms is equivalent to the improbability of drawing the same card twice from a stack of 10^{25} cards stretching from Earth beyond the Andromeda Galaxy. How else can we describe these coincidences except as miracles of fortuity?

Finally, an intriguing aspect of oxygen's presence is that, via its generation of ozone, it indirectly provides protection of all terrestrial life, including the plants that manufacture it, from dangerous UV. This means that oxygen, by providing the protective blanket of ozone in the stratosphere, directly *promotes photosynthesis and thereby its own formation.*

This is not the only example of a key constituent of nature being directly involved in either its own synthesis or preservation. The preservation of water as a liquid in the face of falling air temperatures mainly depends on the fitness of an ensemble of thermal properties of water itself (see *The Wonder of Water,* Chapter 3). Moreover, in the case of oxygen, long before oxygen atoms appeared on Earth, their synthesis in the center of stars depended on the oxygen nucleus and the carbon nucleus having just the right nuclear energy levels. *So in a sense oxygen itself, via its own intrinsic properties,* ensures its presence in nature in two vastly different ways: by its own intrinsic nuclear energy levels in the stars; and by enabling the process of photosynthesis through its offspring ozone on Earth and through its own transparency to visual light!

And ozone does not work alone. There are three independent mechanisms defending biological systems from dangerous UV: (1) The radiant output of the Sun falls dramatically from 0.4 to 0.2 microns so that very little dangerous ultraviolet (far UV) radiation less than 0.2 microns leaves the Sun in the first place; (2) The ozone in the upper atmosphere strongly absorbs radiation of wavelengths below 0.30 microns (from 0.200–0.300 microns[31]); (3) Water (liquid) absorbs strongly from below 0.200 microns. Yet again, it is as if the various players in the game of life, O_2, O_3, and H_2O, were colluding together with the Sun to the same end—protecting life from the harmful effects of UV.

Summary

IN THE article entitled "Electromagnetic Spectrum" in the fifteenth edition of the *Encyclopaedia Britannica,* the authors comment, "Considering the importance of visible sunlight for all aspects of terrestrial life, one cannot help being *awed* by the dramatically narrow window in the atmospheric absorption... and *in the absorption spectrum of water.*"[32]

We should indeed be awed, not just at the "narrow" window in the atmospheric absorption and in the absorption of liquid water, but at the whole ensemble of fitness described in this chapter and the previous chapter (graphically illustrated by Figure 3.7 below) which facilitates the primal process of photosynthesis and hence our own existence.

Clearly the existence of energy-hungry "light eaters" like ourselves, inhabiting the surface of a planet like the Earth, deriving energy generated by the oxidation of the reduced carbons manufactured during the process of photosynthesis, depends on what can only be described as an extraordinarily improbable series of coincidences in the order of things.

And it is also surely worthy of awe that (1) the crucial atmospheric window which lets through the light and makes photosynthesis possible is opened up as if by magic *by the main chemical players in the basic chemical reaction itself*; and (2) the vital heat which gently warms Earth and is an essential ingredient enabling photosynthesis (and all other basic biochemical reactions) is retained in the atmosphere also *by two of the major reactants, H_2O and CO_2*. That these two reactants also *regulate the degree of IR retention* adds to the sheer magic of nature's fitness for complex life on a planet like the Earth! We should indeed be awed that our existence might depend on even tiny details in the fine-tuning, like the IR absorbance window between eight and fourteen microns.

In the face of what appear to be wildly improbable coincidences, one can only repeat Hoyle's comment: "There are no blind forces worth speaking about in nature."[33]

The *teleological web of fitness* and the quite unparalleled parsimony in the compression of so many critical elements of natural fitness within the properties of the four primal gases (O_2, N_2, CO_2, H_2O) is truly extraordinary. It is a purposeful symphony composed of only four atoms—hydrogen, carbon, oxygen, and nitrogen—among the most common atoms in the universe. Yet it is a design transcendent *beyond anything within the mundane realm of our ordinary experience*. Nothing like this wondrous teleology was imagined in even the most spectacular and esoteric visions of any of the great seers or mystics of the past.

The impression of design is overwhelming. Even if a skeptic rejects the design inference, the transcendent web of fitness tells a compelling

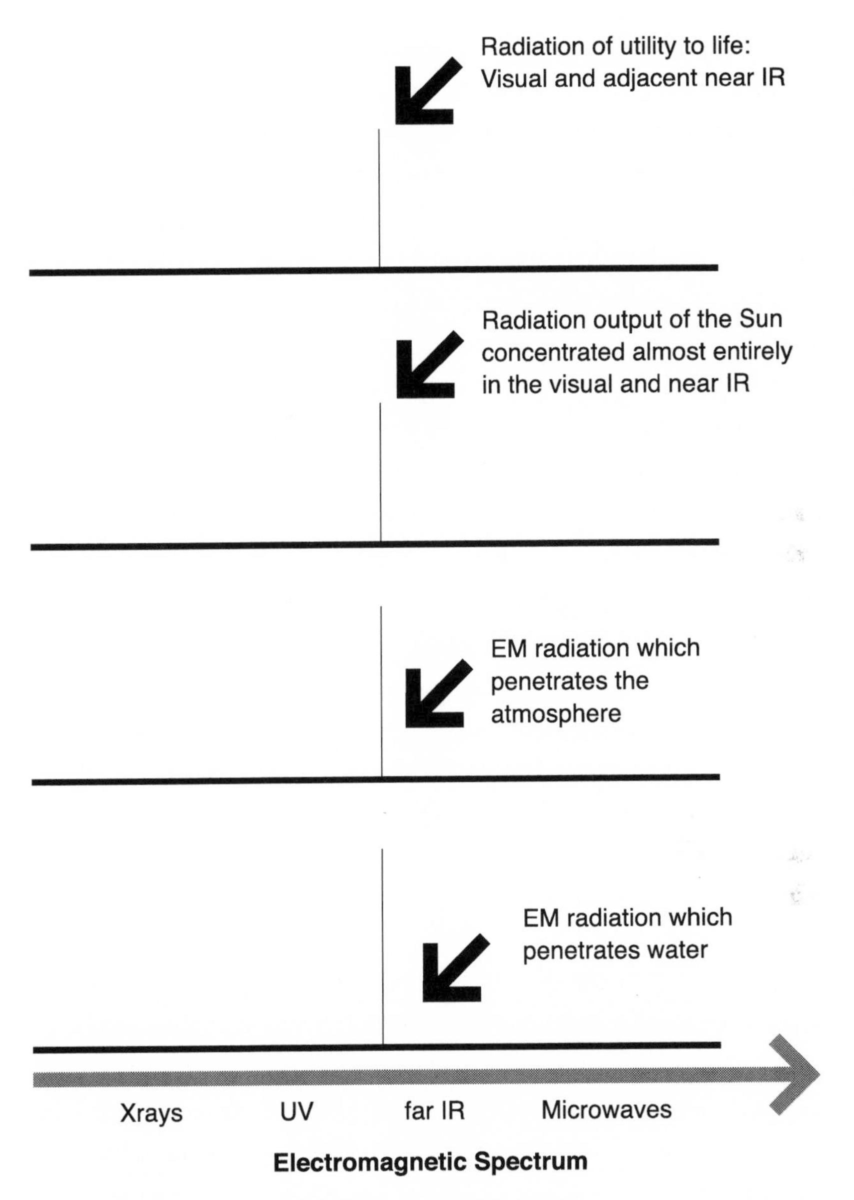

FIGURE 3.7. The narrow windows in the EM that facilitate photosynthesis.

story of a special fitness in nature for oxygen-hungry beings like modern *Homo sapiens,* irrespective of its causal foundation. No other scientific revelation shows so convincingly that the cosmos sings a special song of man and that in a very real sense the human heart beats at the command of the stars.

Moreover, the absurdly improbable coincidences reviewed above are largely irrelevant to the other major domain of carbon-based life on Earth—the vast biomass of "rock eating" denizens of the dark. The fitness of nature for photosynthesis is a fitness for our type of life, a fitness for man! No other body of evidence so thoroughly overturns the Copernican principle.

And this is not the end of the inventory of unique fitness in nature for us air-breathing terrestrial aerobes. Our oxygen-hungry, active lifestyle demands we take oxygen directly from the air via lungs (as opposed to aquatic aerobes, who take dissolved oxygen from water via gills, a topic briefly referred to again in Chapter 4), and this necessitates the terrestrial photosynthesis of plants and trees to supply the reduced carbon fuels for oxidation in the body. As we shall see in Chapter 4, this depends on another extraordinary suite of elements of fitness. The fitness of nature for our terrestrial type of aerobic being is not only proclaimed in the light of the stars and in the properties of the atmosphere which let through the light but also in an additional suite of fitness which enables the existence of terrestrial plants.

We should be awed at these coincidences. Whether or not they are the result of design, whether the universe was created thirteen billion years ago in a Big Bang or the Steady State model holds,[34] whether there is a multiverse or not, whether or not we are alone, they carry the deepest of messages, i.e., *that organisms of our biological design occupy a very special place in nature's order.*

We have seen how the Sun emits the right light for photosynthesis and how the right light penetrates the atmosphere. As we shall see in the next chapter, once sunlight reaches the surface of the Earth, it gifts us, through the work of the leaf, with the vital oxygen which empowers all advanced life on Earth.

Figure 4.1. Tropical foliage.

4. THE GIFT OF THE LEAF

> [W]hen I first entered on and beheld the luxuriant vegetation of Brazil it was realising the visions in the Arabian Nights—The brilliancy of the Scenery throws one into a delirium of delight and a Beetle hunter is not likely soon to awaken from it, when whichever way he turns fresh treasures meet his eye.
>
> —Charles Darwin (1832), Letter to Frederick Watkins[1]

THE SHEER DIVERSITY AND BEAUTY OF BOTANICAL FORMS IS READily apparent in a tropical forest. The leaf forms in a temperate forest are hardly less diverse and striking—compare an oak leaf with a beech or a maple. Even among closely-related maple species (members of the genus *Acer*), there is a spectacular diversity of leaf forms.[2]

But leaves are not just a delight to the senses and an inspiration for poets and artists. All current air-breathing terrestrial animals on Earth, from bumblebees to primates, are ultimately dependent on the reduced carbon fuels or foodstuffs produced by the photosynthetic machinery in terrestrial plants—in blades of grass, in the leaves of a maple, in the fronds of ferns, and in the micro-leaves of the club mosses.

It is true that half of the oxygen used by terrestrial aerobes to burn their reduced carbon fuels is manufactured by algae in the oceans,[3] but terrestrial aerobic life forms like ourselves that generate our energy by oxidizing the reduced carbons produced by photosynthesis are only possible because of the prior existence of terrestrial plants. It is through the work of the leaves of terrestrial plants that the energy of the Sun is captured and gifted to all advanced, air-breathing terrestrial organisms like ourselves.

And it was thus from the beginning of terrestrial life. Green, photosynthesizing plants came first, in the mid- to late-Ordovician[4] some 450 million years ago.[5] They were the essential pioneers that paved the way for the colonization of the land by animals in the mid-Silurian (430 million years ago), including the first centipedes and millipedes, primitive spiders, and the earliest scorpions,[6] followed by the first insects in the early Devonian[7] (400 million years ago), and later by the first terrestrial vertebrates by the mid-Devonian (370 million years ago).

By providing reduced carbon fuels for land-based life, the gift of the leaf had the enormous consequence of enabling aerobic life forms not only to leave the water, but to become air-breathing—taking up oxygen directly from the atmosphere. Previously, all complex aerobic organisms (fish and invertebrates) had been aquatic and breathed through gills, taking up their oxygen from the water. We *Homo sapiens*, for example, are not just obligate *aerobes*; we are obligate *air-breathing aerobes, deriving our oxygen directly in gaseous form from an atmosphere.* Only by taking in oxygen directly from an atmosphere enriched in oxygen (as is our current atmosphere on Earth) can we obtain the necessary 250 milliliters of oxygen we need every minute even at rest.[8]

And there is little doubt that this requirement (being air-breathing) will also apply to all advanced, complex carbon-based aerobes throughout the universe. This is because of a fundamental constraint. It is far more difficult to obtain oxygen from water than from air, and this puts a ceiling on the metabolic rate aerobic water-breathing organisms can attain and on the consequent complexity (in the broadest sense) that aquatic organisms may achieve compared with air-breathing organisms.[9] And this almost certainly precludes the development in aquatic organisms of large brains and high intelligence.

Whereas the use of oxidation, as George Wald pointed out, provided living things with extra capital over and above that provided by anaerobic metabolism and allowed them to grow in complexity and do interesting things beyond mere surviving,[10] *air-breathing* (to paraphrase Wald[11]) provided even more capital to be even *more complex* and do even more interesting things.

Because of the *universal* and absolute requirement for oxidation (of reduced carbon) to provide sufficient metabolic energy for advanced life forms; because of the additional constraint, the need to extract the oxygen from an atmosphere to reach the high metabolic rates possessed by all the most intelligent life forms on Earth; and because of the necessity for a terrestrial environment for the mastery of fire and the development of technology and an advanced technological civilization, no intelligent life forms and certainly none that possess an advanced technology ever have breathed or ever will breathe via gills and live in the sea.

No matter that the Sun and the vast majority of stars beam out that tiny fraction of the EM spectrum that is so supremely fit for life (light and heat), and no matter that the atmosphere has the right spectral absorbance properties to let through the light and absorb a fraction of the IR, warming the Earth and animating the stuff of life for chemical reaction. All these remarkable elements of fitness would be of no avail for us terrestrial "light eaters" if we could not utilize the energy of the Sun for either respiration or combustion by oxidizing the reduced carbon fuel supplied by green plants.

Without the gift of the leaf there would be no herds of wildebeest, zebra, and antelope grazing the African Savannah, no bumblebees drinking nectar from a flower, no fire, no metallurgy, no chemistry, no technology, and no humans planning a trip to Mars.

FIGURE 4.2. *Bombus barbutellus* (a species of the cuckoo bumblebee) enjoying the gift of the leaf.

Photosynthesis

ALL THE oxygen we breathe and the vast majority of oxygen that has ever existed on Earth since the formation of our planet has been generated by the process termed *oxygenic photosynthesis*, energized by photons of light beamed to Earth from our Sun. The process first began in the primeval oceans some 3.5 billion years ago, carried out by primitive blue-green algae (cyanobacteria), later by various types of marine algae, and finally, over the past four hundred million years, by terrestrial plants.

And not only the oxygen but all the food—the reduced carbon compounds which we aerobes oxidize to generate our metabolic energy—is also ultimately the product of photosynthesis. In our case, and in the case of all other terrestrial aerobes, it is given to us through photosynthesis carried out in the leaves of terrestrial plants and trees.

The global scale of the process of photosynthesis is immense, fixing three hundred trillion kilograms of CO_2 per year[12] and liberating two hundred trillion kilograms of oxygen.[13]

In essence, the process involves the use of light energy to draw electrons and protons (H+) from water (H_2O), oxidizing the water to oxygen (O_2) which is released into the atmosphere, and reducing carbon dioxide to sugars and various reduced carbon compounds (CH). The overall reaction can be written thus:

$$CO_2 + H_2O \Rightarrow CH + O_2$$

I am using the abstract formula CH in this section of the text to stand for all reduced carbon compounds of the general formula $C_xH_y(O)_z(N)_w$ (e.g., glucose, $C_6H_{12}O_6$), which make up the material substance of organisms and provide the fuel for respiration (food) and combustion (wood); all such compounds contain atoms in addition to carbon (C) and hydrogen (H), atoms such as oxygen (O) and nitrogen (N).

We aerobes then oxidize the reduced carbon compounds manufactured during photosynthesis, generating metabolic energy and releasing CO_2 and H_2O into the atmosphere. This is why Broad could call us "light eaters," because we do indeed live off of, or eat, the proceeds of

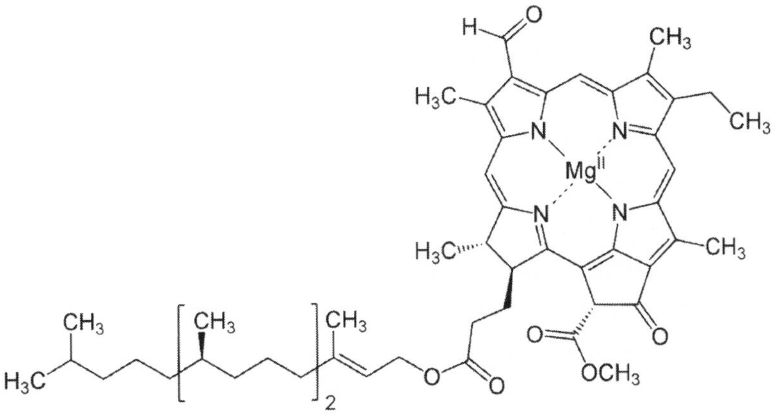

FIGURE 4.3. Chlorophyll is the most important light-harvesting pigment on earth, present in all photosynthetic organisms.

photosynthesis. We terrestrial aerobes obtain our vital foodstuffs from photosynthesis carried by terrestrial plants. We are indeed "light eaters," entirely dependent on the gift of the leaf.

"Light eating" is not the *only* way of life for carbon-based organisms on Earth. Curiously from our oxygen-centric perspective, the majority of life forms on Earth, and almost certainly more than half the biomass, live without the Sun and have no need for either the oxygen or the reduced carbon fuels generated by photosynthesis (see Appendix A, "Doing Without Sunlight").

A complete description of photosynthesis is beyond the scope of this chapter. Excellent descriptions of the details are available in every major biology text. But omitting a million details and greatly simplifying things, including chemical formulae, oxygenic photosynthesis proceeds as follows:

The primary event on which the whole process of photosynthesis depends is the capture or absorption of photons of light by the photosynthetic pigments (chiefly the green pigment chlorophyll; Figure 4.3) in the thylakoid membranes (which surround the so-called thylakoid discs in the chloroplast—see Figure 4.4). When the chlorophyll molecules situated in these membranes capture photons, the energy imparted activates electrons in the chlorophyll, raising them to higher energy levels. (Each photon absorbed raises one electron to a higher energy level).[14]

This allows the electrons to escape from the chlorophyll, leaving the chlorophyll molecules positively charged or oxidized. (The loss of electrons is oxidation.) The positively charged chlorophylls draw electrons from water molecules (H_2O) in the oxygen-evolving complex (OEC), oxidizing them and releasing at the same time free oxygen (O_2) molecules, as well as protons (H+) and electrons (e-).

Water [H_2O] ⇨ Oxygen [O_2] + protons [H+] + electrons [e-]

The energetic electrons escaping from the chlorophyll find their way to electron transport chains, where they flow "down" in discrete steps, releasing energy at each step, which is used to do work, pumping protons (H+) across a membrane (the thylakoid membrane) into the thylakoid lumen (a membrane-enclosed compartment in the chloroplast). These then flow back through the same membrane, providing energy to drive the synthesis of ATP (the cell's chemical energy currency) by the enzyme ATP synthase.[15]

STROMA			
	H+		ATP
THYLAKOID MEMBRANE	⇩		⇧
LUMEN	H+	⇨	H+

Thus, *solar energy* is converted into *chemical energy*, which is subsequently used by the cell in the reduction of CO_2.

In a second electron transport chain, the excited electrons flowing along the chain also generate energy to pump protons across the thylakoid membrane generating ATP (as above) and to generate a compound whose acronym is NAPDH (which acts as a store of reducing equivalents).

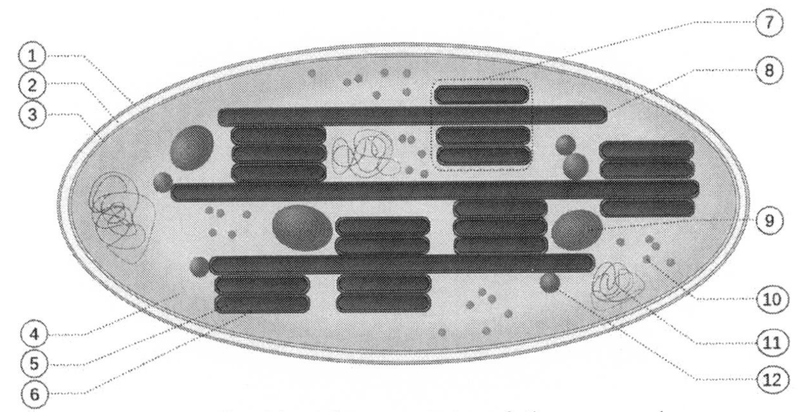

FIGURE 4.4. The chloroplast, consisting of: 1) outer membrane; 2) intermembrane space; 3) inner membrane (1+2+3: envelope); 4) stroma (aqueous fluid); 5) thylakoid lumen (inside of thylakoid disc); 6) thylakoid membrane surrounding disc; 7) granum (stack of thylakoids); 8) thylakoid (lamella); 9) starch; 10) ribosome; 11) plastidial DNA; 12) plastoglobule (drop of lipids).

The ATP and NAPDH are then used to reduce carbon dioxide, creating reduced carbon compounds (carbohydrates) in the following greatly simplified reaction:

$$CO_2 + ATP + NADPH \Rightarrow CH + ADP + NADP$$
$$\text{Carbon dioxide} + \text{energy} + H \Rightarrow \text{Reduced carbon}$$

Overall, photosynthesis can be seen to occur in two stages. In the first stage, *light-dependent reactions* capture the energy of light and use it to make the energy-storage molecule ATP and the reducing agent NADPH. These light-dependent reactions occur in the thylakoid membranes. During the second stage (the Calvin cycle; see Figure 4.5), the *light-independent* reactions use these products to reduce carbon dioxide.[16] These light-independent reactions occur in the stroma of the chloroplast.

Overall, the energy captured from light energizes the transfer of matter from water, in the form of electrons and protons to carbon dioxide (CO_2). This forms reduced carbon compounds (CH) and releases free oxygen (O_2) as a byproduct—thus transducing light energy into chemi-

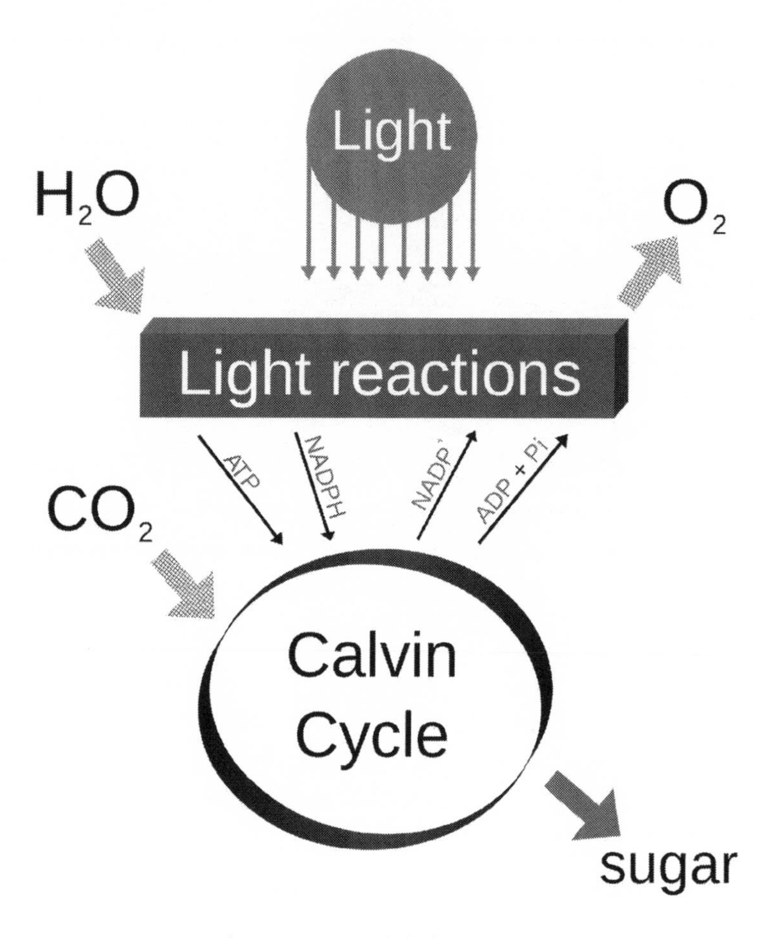

FIGURE 4.5. Light reactions and dark reactions (Calvin Cycle).

cal energy, stored in the reduced carbon compounds upon which we feed (i.e., which we oxidize) to obtain metabolic energy.

Oxygenic photosynthesis is not the only type of photosynthesis. Many microbes practice *non-oxygenic photosynthesis*—the exploitation of light as an energy source for biosynthesis without the generation of O_2.[17] Small amounts of oxygen are also produced by certain microbes, by processes powered by metabolism and not sunlight.[18] Intriguingly, humans are making progress towards the goal of non-biological, artificial, oxygenic, photosynthetic systems as a route to sustainable clean energy.

Some researchers are attempting to mimic biological oxygen production, while others are using chemical methods, which are quite different from those used in the chloroplast.[19]

However, at present, the only known mechanism for producing the copious quantities of oxygen needed to generate an aerobic atmosphere like that on the Earth is the kind that occurs in the chloroplasts of green plants and in the cells of blue-green algae (cyanobacteria). In nearly four billion years of cellular evolution, no other mechanism for oxidizing water and releasing oxygen has evolved. The oxygen-evolving complex system is essentially identical in all oxygenic photosynthetic organisms, from giant redwoods to blue-green algae, and is presumed to have arisen only once in the history of life on Earth.[20]

Photosynthesis is not the only example of photochemistry utilized by biological systems. There are many other examples. One, of particular relevance to us humans, occurs in the photoreceptors of the vertebrate eye, which are discussed in Chapter 5 below. Another photochemical reaction is the synthesis of Vitamin D_3 in the skin from steroid precursors by the action of sunlight in the near UV range (between 0.29 and 0.32 microns).[21]

And photochemistry is not restricted to biological systems—it has many industrial applications. One of the simplest and best-known is the use of light-sensitive emulsions in pre-digital photography. Photons react with crystals of silver chloride, exciting the electrons in the chloride ions (analogous to their excitation in chlorophyll), which subsequently jump to the adjacent silver ions, neutralizing them and converting the white crystals of silver chloride to black particles of elemental silver and chlorine gas, generating a negative image on the photographic plate.

The polymerization of polymers by UV light is another industrial application of photochemistry[22] as is the rapid hardening of dental putty using blue light.[23]

Water and the Leaf

Despite the fitness of sunlight for photosynthesis and the fitness of the atmosphere to let through the right light, photosynthesis in land plants is dependent on many additional elements of fitness. We have already seen that water is a key player in photosynthesis, providing the electrons and protons used in the reduction of carbon dioxide and the waste product oxygen. Water also has just the right absorbance properties as a vapor and a liquid to let through the *right light* and absorb some of the IR radiation to warm the Earth. Without this suite of properties, there would be no photosynthesis in aquatic or terrestrial plants and no vital transduction of the Sun's beneficence for land-based, air-breathing aerobes.

But even more remarkably, water possesses another suite of physical and chemical properties which are specifically fit for terrestrial plants and the vital greening of the land.[24] All the elements of fitness in nature for photosynthesis would be of no utility for us terrestrial aerobes if water were not fit in intriguing ways that enable leaves of terrestrial plants to provide the basic reduced carbon nutrients for *Homo sapiens* and all other terrestrial air-breathing species. No consideration of the fitness of the light would be complete without some discussion of water's fitness for photosynthesis in terrestrial plants.[25]

The Hydrological Cycle. It is well-known that water evaporates from the sea, rises into the atmosphere, cools, and eventually condenses into tiny droplets, forming clouds. These coalesce into larger droplets which eventually fall to the ground as rain or snow. Water from rain and melting snow drains into rivers and is carried to the sea. Ice is carried in glaciers (in higher latitudes) to the sea. The scale of this remarkable cycle is massive, as Ball points out:

> Each 3100 years, a volume of water equivalent to all the oceans passes through the atmosphere, carried there by evaporation and removed by precipitation… the Sun's heat removes from the oceans the equivalent of a one-metre depth each year—875 cubic kilometres in total every day.[26]

This cycle depends on a unique property of water that is hardly ever acknowledged. As Ball comments: "This cycle of evaporation and condensation [which empowers the hydrological cycle] has come to seem so perfectly natural that we never think to remark on why no other substances display such transformations."[27]

But as Ball comments further: "The very existence of a hydrological cycle is a consequence of water's unique ability to exist in more than one physical state—solid, liquid or gas—under the conditions that prevail at the surface of the planet."[28] And as he points out: "Almost all of the non-aqueous fabric of our planet remains in the same physical state. The oxygen and nitrogen of the air do not condense; the rock, sands, and soils do not melt... or evaporate."[29]

Only water fails to conform. *Of all known substances, only one—water*—exists in the three material states in the ambient temperature range and is fit for the hydrological cycle and its major consequence: the delivery of water to land-based life. Although the importance of the hydrological cycle is widely acknowledged, what is rarely or never mentioned is that the delivery of water to the land is in effect carried out by and dependent on the properties of water itself, unaided by any other external regulatory systems. Without the hydrological cycle, the entire land surface of Earth would be a dehydrated lifeless waste, more lifeless than the Atacama or any of the most dehydrated deserts currently existing. The light beaming down would be useless to us, because no plants would green the land and turn the light into oxygen for us energy-hungry, terrestrial aerobes.

Erosion and Weathering. While the hydrological cycle delivers an endless supply of water to the land, making terrestrial life and ecosystems possible, and this might perhaps be considered its primary function, it does much more than this. Inevitably as the wheel turns, it performs another major task critical for terrestrial plants. The tumbling waters of a million mountain streams coursing over the exposed lithosphere continually leach minerals from the rocks and enrich and replenish all the waters of the terrestrial hydrosphere, including lakes, rivers, subsurface ground waters, soil, wet lands, etc., with the vital elements of life which are essential to all living things on Earth. Thus the hydrological cycle achieves another aim essential to plants: It leaches minerals from the rocks and by percolating continually through the soils and ground waters, distributes them to all parts of the terrestrial hydrosphere.

The weathering and solubilization of the rocks by water is greatly enhanced by the fact that carbon dioxide is soluble in water and reacts with water to produce a mild acid, carbonic acid, which aids in the solubilization of the minerals in the rocks and the release of vital elements into the terrestrial hydrosphere.

The erosion and weathering process is further promoted by two other unique properties of water which play an important role in erosion and weathering—water's high surface tension (the second-highest, after mercury, of any common liquid[30]) and its anomalous expansion on freezing. Working together, these properties cause the fracturing of rock as water, which is drawn into fissures in the rocks because of its high surface tension, subsequently expands on freezing. The expansion is about ten percent in terms of volume and exerts tremendous pressure on the surrounding rocks, breaking them down and presenting a greater surface area for chemical weathering. The expansion of water on freezing is again, like its existence in three states in ambient conditions (solid, liquid, and gas), another anomalous and practically unique property of water. Only one other known substance, the metal gallium, expands on freezing in the ambient temperature range (26°C).[31]

The low viscosity of liquid water and solid water (ice) are also critical elements of fitness which contribute further to the erosional work of water and to the subsequent chemical weathering of the rocks. Water's low viscosity and mobility, in conjunction with the abrasive action of the tiny particles of rock carried within the stream, greatly enhances water's erosional powers. Additionally, the viscosity of ice is very low for a crystalline solid, only about 10^{11} Pa-s,[32] about ten orders of magnitude lower than that of the rocks in the Earth's crust. If the viscosity of ice had been closer to the viscosity of rock, all the waters of Earth would be locked up in vast immobile ice caps at the poles and in high mountain regions, but because of its relatively low viscosity mountain glaciers and the great continental ice sheets eventually flow back to the sea.

And as the glaciers flow, sliding over the bedrock, they drag rocks and rock fragments across the underlying surface, grinding away the underlying rocks and reducing them to "rock flour," material made up of tiny grains of rock, fractions of a millimeter in size. This greatly increases the surface area of material available for chemical weathering.

Thus, water's fitness for its role in eroding and weathering the rocks is due to a remarkable synergy in a suite of diverse properties that act together to achieve the absolutely vital end of eroding and weathering the rocks and providing terrestrial plant life with the necessary vital elements and nutrients essential for their growth. And this synergy is in turn only possible because of the prior fitness of water to enable the hydrological cycle.

Without this, there would be no plants on Earth to carry out photosynthesis because they would not have the nutrients needed to grow, and the life-giving work of the Sun would be inaccessible to us.

Soil. While the hydrological cycle provides both water and the necessary minerals for terrestrial life, there is still another condition that must be satisfied if land-based plants are to thrive. Water, enriched with the vital nutrients of life through erosion, must be retained in the soil, to provide an accessible and long-term supply of water and nutrients for absorption by the roots of land plants. Amazingly, the inevitable end of water's work in eroding and weathering the rocks *also* results in the formation of a variety of mineral components that together make up an ideal matrix for retaining water and its cargo of dissolved mineral nutrients.

This matrix includes a variety of sands (grain size greater than sixty-three microns) and silts (between two and sixty-three microns) and various types of clay (particle size under two microns), which comes from the chemical weathering of silicates. This mix of non-organic matter in the soils provides a vast surface area and a labyrinth of micropores in which water is retained by capillary action. Such water is often termed *capillary water* in contrast to *gravitational water*. The latter resides in macro-pores and can drain rapidly from the soil.

The water-retaining property of soils is crucial to the survival of plants. As leading soil scientists Nyle Brady and Raymond Weil comment:

> As long as plant leaves are exposed to sunlight, the plant requires a continuous stream of water to use in cooling, nutrient transport, turgor maintenance, and photosynthesis. Because plants use water continuously, but in most places it rains only occasionally, the water-holding capacity of soils is essential for plant survival. A deep soil may store enough water to allow plants to survive long periods without rain.[33] [Acacia trees on the African savannah may go without water for almost five months.[34]]

It is easy to imagine the dire consequences for plant life if soils only contained macropores through which gravitational water could drain quickly away, or alternately if soils were impermeable. After one week without rain, the majority of plants would begin to wither and die. Over time, the result would be the de-vegetation and desertification of the entire Earth's land surface.

There is a beautiful and elegant teleology in all this. The same process both draws minerals and nutrients from rocks and generates the clays and sands and silts that together form soil, which, along with organic debris, forms an ideal water and mineral-retaining matrix for terrestrial plants.[35]

Reaching the leaves. The unique properties of water have another vital but seldom acknowledged role in terrestrial photosynthesis. They allow water be drawn up the stems of plants and trunks of trees to the "business end" of terrestrial plants—the leaves where the chemical reactions of photosynthesis occur.[36]

Steven Vogel, in his book *The Life of a Leaf*, describes the way water manages to get to the top of tall trees as a tale *mirabile dictu* (wonderful to relate).[37] It turns out that this feat depends not only on the high surface tension of water but also on another unusual property of water, its remarkable and counterintuitive *tensile strength*, which like its high surface tension is also the result of its hydrogen-bonded network.

Simple capillarity, caused by surface tension (a generic property of all fluids), can easily lift water up to one hundred meters if the tube is small enough. In tubes one hundredth of a micrometer (ten nanometers) in diameter, the surface tension is so strong that it can support a column of water three kilometers or two miles high;[38] however, because of viscosity (a measurement of internal friction), water's resistance to flowing through such tiny conduits would be prohibitively high.[39] The conduits in trees are between 0.03 and 0.3 millimeters in diameter, sufficiently wide to allow the water to flow up through the tubes with minimal resistance. But as Vogel comments, "Thirty micrometers sends water only about 1.5 meters (5 feet) upward, and 300 micrometers is ten times worse: 15 centimeters, or 6 inches."[40]

So how do trees manage to exploit capillarity to hold a column of water one hundred meters high (requiring tiny tubes) while at the same time overcoming viscous drag? As Michele Holbrook and Maciej Zwieniecki explain, plants solve the problem of the viscous drag "by connecting the small capillaries in leaves [which are small enough to generate capillary forces powerful enough to hold a column 100 meters high] to larger ones that provide a much wider transport channel that runs from the veins in the leaf down through the stem and into the roots."[41]

The key point is that the critical capillary forces are not generated in the major conduits. As Holbrook and Zwieniecki point out:

> The relevant capillary dimensions are not those of the relatively large conduits that you would see if you cut down a tree and looked inside... (diameters of 50–100 μm). Rather, the appropriate dimensions are determined by the air-water interfaces in the cell walls of the leaves, where the matrix of cellulose microfibrils is highly wettable and the spacing between them results in effective pore diameters [which function as tiny capillaries] of something like 5 to 10 nm.[42]

And this is the crucial point: The pore diameter is so small that the surface tension generated (as mentioned above) is able to support a water column three kilometers high, much higher than the tallest tree.

In other words, as the authors continue:

> Trees and other plants overcome... [the problem] by generating capillary forces in small-diameter pores [at the interfaces in the leaves between the sap and the air] but transporting water between soil and leaves through larger-diameter conduits. That strategy allows them to achieve greater heights than with a straight-walled microcapillary.[43]

It is worth noting that the low viscosity of water again plays a vital role. If the viscosity were much higher, the conduits between root and leaf might have to be wider and take up an incommensurate volume of the trunk!

But while capillarity suffices to hold up the one-hundred-meter column, what pulls the water up from the roots through the conduits to the stems and leaves at the top of the tree? The answer is that the evaporation or transpiration from the air-water interfaces in the leaf cell produces suction by inducing a negative pressure in the fluid under the tiny menisci, which is transmitted to the whole system of conduits. It is a basic law of hydraulics that pressure in one part of an enclosed hydraulic system is transmitted to all other parts. As water molecules are lost from the leaves at the top of the tree, others must enter the roots to take their place. The continual loss of water molecules lowers what is termed the *water potential* by increasing the solute concentration in the regions below the interfaces. (Water moves from a region of high water potential to a region of low potential. Increased solute concentration lowers the water potential. Increased pressure increases it.) This lowering of potential, transmitted to the whole hydraulic network, pulls the water up the conduits to the interfaces where it is lost by evaporation to the atmosphere.

Another question arises: Why does the column of water not break into pieces as it is tugged from above? The answer lies in the cohesiveness of liquids, more pronounced in water than most other common fluids because of water's hydrogen-bonded network, which holds water molecules together! And because of this property of water columns, although the notion is very counterintuitive, water in the conduits has tensile strength.[44]

Tensile strength is the ability of a substance to resist being stretched. A steel wire will resist breaking on being stretched because of the tensile strength of steel. Curiously, it is the same with a water column. As Vogel points out, experiments show that a rope of liquid water, a square centimeter in cross section in an enclosed tube, has sufficient tensile strength to support a solid mass of nearly three hundred kilograms.[45] And as Vogel points out, steel is stronger, but only by ten times! It is this very counterintuitive tensile strength of fluids—especially water—that allows the negative pressure caused by the evaporation in the leaves to pull water from the roots one hundred meters up to the leaves without any breakage in the column.

This fascinating story of how water ascends from soil to leaf, so vital to the existence of large trees, depends critically on two basic physical properties of water as a fluid: its tensile strength, which means the "pull of evaporation" will not break the water column, and the enormous surface tension generated by water in very narrow tubes or passages in the leaf.

The mechanism represents a unique and stunningly brilliant solution to the problem of raising water to the top of large trees. Significantly, no conceivable alternative could work. Vogel in his *The Life of a Leaf* waxes lyrical in contemplating the way it's done: "The pumping system has no moving parts, costs the plant no metabolic energy, moves more water than all the circulatory systems of animals combined, does so against far higher resistance, and depends on a mechanism with *no close analog in human technology*."[46] Holbrook and Zwieniecki concur:

> Trees can be rightly called the masters of microfluidics. In the stem of a large tree, the number of interconnected water transport conduits can exceed hundreds of millions, and their total length can be greater than several hundred kilometers. Furthermore, on a sunny

> day, a tree can transport hundreds of gallons of water from the soil to its leaves, and apparently do it effortlessly, without making a sound and without using any moving parts… The physics that underlies water transport through plants is not exotic; rather, the application of that physics in the microfluidic wood matrix results in transport regimes operating *far outside our day-to-day experience.*[47]

Of course, trees are only possible because of an ensemble of additional elements of fitness in nature, in addition to the unique mechanism described above. These include the unique properties of the cellulose-lignin composite, which confers tensile strength and durability to tree trunks and also promotes the formation of soil.[48] Additionally, water's power evaporative cooling protects leaves from overheating in the Sun.[49] And of course trees depend on the water-retaining properties of soil.

Summary

Water is not only one of the key chemical reactants of photosynthesis, but, as we saw in Chapter 3, as a vapor and a liquid it has just the right absorbance properties to let through the visual light and retain some of the heat emitted by the Sun. Now we see that another suite of physical properties are uniquely fit to enable the chemical process of photosynthesis to occur in the leaves of land plants, permitting the greening of the land and the vital growth of large trees, providing foodstuffs for energy-hungry aerobes like ourselves and fuel for the making of fire.

Complex terrestrial life relies ultimately on the energy of the Sun, but the life-giving nature of light is transmitted to terrestrial aerobes through the work of the leaf. But the gift of the Sun and the work of the leaf can only be of utility to terrestrial aerobes like ourselves because of the unique fitness of water for terrestrial plants.

Next, we will turn to the most familiar way that light is specifically fit for us, through the gift of vision.

FIGURE 5.1. The miracle of vision.

5. Fitness for Vision

All men by nature desire to know. An indication of this is the delight we take in our senses; for even apart from their usefulness they are loved for themselves; and above all others the sense of sight. For not only with a view to action, but even when we are not going to do anything, we prefer sight to almost everything else. The reason is that this, most of all senses, makes us know and brings to light many differences between things.

—Aristotle (mid-fourth century B.C.), *Metaphysics*[1]

There is no doubt Aristotle is right in his assessment of vision, as seen in the passage at the head of this chapter. Sight does indeed "bring to light many differences between things." To make a fire, to smelt metals, to do chemistry, you need to be able to see. Indeed, without vision, more specifically high-acuity vision, our knowledge of the external world would be vastly impoverished and it is doubtful whether any carbon-based aerobic life form anywhere in the cosmos, however advanced and intelligent, would be able to develop a sophisticated technology and come to have some understanding of nature or its place in the cosmos.

Some organisms, including bats and dolphins, can use sound to provide a remarkably detailed image of their environment through echolocation. Bats use it to successfully navigate through thick vegetation in pitch darkness to hunt and catch flying insects.[2] Even some humans who are completely blind can learn to master the technique of using sound to navigate, enabling them to play football, basketball, and even ride a bicycle through traffic.[3] The sense of touch shared by all organisms—taken to extraordinary length in the case of the star-nosed mole which has 25,000 touch sensors on its nose, many more in an area the size of a human fingertip than in the whole human hand[4]—can also bring us much essential information about our immediate environment. And in some organisms, like dogs and bears, the sense of smell is very highly developed.

But while sensing the environment using echolocation may enable many organisms (even humans) to navigate in the dark with extraordinary efficiency and a great deal of information regarding our immediate environment can be gained by the senses of touch and smell, no organism restricted to the use of sound, touch, or smell would ever see the Moon, witness a lightning strike, observe a *Paramecium* down a microscope, or watch the dancing flames of a campfire and the glowing embers as the fire slowly dies in the night. It is very doubtful that blind organisms restricted to sensing their environment through sound, touch, and smell would ever master fire and develop metallurgy and an advanced technology.

From the beginning, it was seeing the stars and planets and recording their regularities that spurred the development of science.[5] For beings lacking eyes, however intelligent, fire would have been a mysterious and frightening demon, its mastery beyond control, and the route to metallurgy and chemistry and advanced technology would be forever foreclosed.

The Right Energy

The diversity of biological devices for detecting light is spectacular, ranging from simple eye-spots in unicellular organisms to the high-acuity compound eyes of the dragonfly and the high-acuity camera eye of humans and other vertebrates.[6] All biological light-detecting devices depend on the fundamental fact that the energy levels of EM radiation in the visual region are just right for photochemistry. As discussed in Chapter 2, light is the only type of electromagnetic radiation that has the appropriate energy level for interaction with and detection by biological systems. UV, X-ray, and gamma-ray photons are too energetic and highly destructive, while photons in the infrared, microwave, and radio regions lack the energy necessary to activate bio-molecules for chemical reactions and cannot be detected by bio-systems.

It is worth recalling again that the visual region represents the tiniest fraction of the entire span of the EM spectrum. Recall the diagram emphasizing the smallness of the visual region in the vast range of the EM spectrum.

We are fortunate indeed that such a tiny, useful region of EM radiation exists and doubly fortunate, as discussed in Chapter 3, that radiation in this same small region penetrates both the atmosphere and liquid water. In the human eye, for example, the light must penetrate approximately 2.5 centimeters of aqueous fluid between the front of the eye and

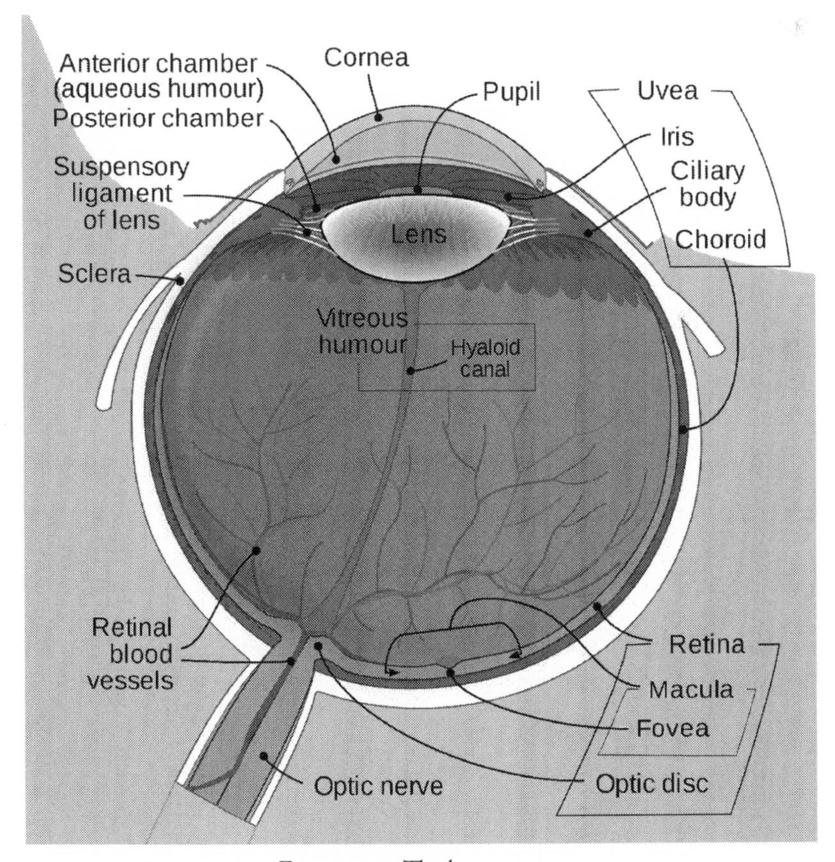

FIGURE 5.2. The human eye.

the retina, including the lens and cornea, which are mainly composed of water.[7] If water absorbed light in the visual regions, both vision and photosynthesis would be impossible.

But these two conditions, that EM radiation in the visual region has *the right energy levels* for bio-detection and that both the atmosphere and liquid water let through *the right light,* while necessary for high-acuity vision in carbon-based life forms like ourselves, are not sufficient. The high-resolution, camera-type eye found in all higher vertebrate species, including man, is only possible because of several additional elements of fitness, including most importantly the actual *wavelength of light itself in the visual region.*

Remarkably, not only is the energy of the photons in the visual region *right for bio-detection;* the wavelength in the visual region is also *right for the formation of an image in a high-acuity eye* of a size and design commensurate with biological organisms of our size and design.

High-acuity Eyes

Among vertebrates all high-acuity biological eyes, including our own, are single-chamber, camera-type, image-forming devices.[8] All possess a lens that focuses an image of objects in the environment onto a light-detecting organ, a retina. All detect incident light by special light-detecting cells—photoreceptors—which are all basically long, narrow, tubular structures ranging in diameter from one to a few microns in diameter. High-acuity camera-type eyes are found only in vertebrates and cephalopods.

In all photoreceptors, there is a specialized region packed with photoreceptor molecules. For humans these molecules are contained in the "outer segments" of our photoreceptors, which are about twenty-five microns long and two microns wide in the case of the rods (the dim light receptors in all vertebrates) and about thirteen microns long and tapering from three microns wide at the base to one micron at the tip in the case of the cones (bright light receptors).[9] The photoreceptor cells in all high-acuity eyes are packed with millions of photon-detecting molecules—150,000,000 in the case of human rods.[10] The narrow tubes act as wave guides.[11] Tiny optic fibers trap the light rays by internal reflection and force the photons along the tube through the forest of receptor molecules.[12]

A detailed description of the photoreceptors in the human retina and the mechanism by which they capture photons is given in all major texts of ophthalmology, including those consulted here.[13]

The basic design of all high-acuity eyes—consisting of a lens, a retina, and long narrow photoreceptor tubes crammed full of photon-detecting molecules—is deducible from first principles, so it is no surprise that the same design occurs in organisms as diverse as an octopus and a human! Although the retina of cephalopods is not inverted as in the vertebrate eye, all the major features of the vertebrate camera eye are present—a lens, a retina, and narrow tube-like photoreceptors. And the photoreceptors in the octopus eye are similar in basic size and shape to those in vertebrates: about sixty microns long and about three microns in diameter in retinal regions where the resolution is highest.[14]

The need for a lens. While a simple pinhole-camera-type eye lacking a lens (like that of the marine cephalopod *Nautilus*) will focus an image without a lens, the resolution is far less—only a fraction of that achieved in eyes with lenses—and the brightness of the image is also greatly diminished because of the very small diameter of the pinhole necessary to provide even a poorly resolved image.[15]

The necessity of narrow photodetecting tubes. Narrow photodetecting tubes allow a huge number to be packed side by side in the retina, up to 200,000 per square millimeter in regions of the retina in man[16] and 1,000,000 per square millimeter in eagles and hawks,[17] giving the raptors very high pixel density and high resolving power (the highest known of any high-acuity eye on Earth).

The number of individual photoreceptor molecules per tube. The extremely high number of photoreceptor molecules in each tube confers sensitivity on the retina by enabling the rod photoreceptors (adapted for night-time vision) to respond to a very low photon flux and see in dim light. Remarkably, individual rods can capture and respond to an individual photon of light, the minimum parcel of light energy. As Michael Land comments in the *Encyclopaedia Britannica*:

> Rods operate over the range from the threshold of vision, when they are receiving about one photon every 85 minutes, to dawn and dusk conditions, when they receive about 100 photons per second. For most of their range the rods are *signaling single photon captures.*[18]

Clearly the more photoreceptor molecules a photoreceptor cell contains, the greater the chance of photon capture. As Yingbin Fu comments:

> The *dense stack of discs* of the rod outer segment ensures that *virtually every photon traveling axially will be captured.* In a sense, vertebrate rods can be viewed as sophisticated three-dimensional photon capture devices.[19]

And despite the 150,000,000 photoreceptor molecules stacked in the outer segments of the rods, *some* photons do escape, evidenced by the fact that many nocturnal animals, cats and owls for example, have a reflecting membrane, the *tapetum lucidum,* at the back of the retina to reflect back any photons which evade capture in its first pass through the photoreceptor.[20] (That's why, in New Zealand and Australia, we call the reflectors in the middle of the road "cat's eyes"). Any decrease in the number of photoreceptor molecules packed into the detector tubes would lead to a loss of sensitivity.

The ability of the rod photoreceptors to generate a neural signal after detection of an individual photon is only possible because of an intricate amplification system which magnifies the initial chemical change in a single photoreceptor molecule about one million times.[21]

Photoreceptor cells not only need to capture and respond to an individual photon when called upon to see in the dark; they also need to be able to function in the daytime when, in some environments, the luminosity is much greater. (Amazingly, the eye can see on a dark night and on a brightly illuminated ski field spanning a difference in luminescence on the order of 10^{12} or 1,000,000,000,000.[22])

The high number of photoreceptor molecules packed inside the cones (the daylight photoreceptor cells) allows the cones to function in very high photon fluxes (i.e., to see in bright light). As Michael Land comments:

> The cones are much less sensitive than the rods; they still respond to *single photons*, but the sizes of the resulting electrical signals are much smaller. This gives the cones a much larger working range, from a minimum of about three photons per second to more than *a million per second*, which is enough to deal with the brightest conditions that humans encounter.[23]

In bright daylight—on a ski field, for example—the flux of photons impinging on an individual photoreceptor is so great that seeing requires a vast number of receptor molecules in the cones *permanently ready to respond to an individual photon* so that the photoreceptor cells can be continually responsive all the time to changing high levels of incident light entering the eye.[24] But for the cones to capture up to one million photons per second, allowing us to see in sunlight, there must be *one million photoreceptor molecules*, available for charging every second and ready to respond to a photon of light.

Studies show that it takes about three to four minutes[25] to recharge all the photoreceptor molecules in a cone, but the majority are recharged in about two minutes. Because there are on the order of 100,000,000 photoreceptor molecules in an individual cone, given a recharging time of approximately 100 seconds, *1,000,000 charged receptors will be available for photon detection every second, ensuring the continuance of vision even in bright light!*

Seeing both by moonlight and on a ski field, where the difference in luminescence is on the order of a trillion-fold, depends on a vast number of photoreceptor molecules in each kind of photoreceptor cell.

Diffraction

Although our eyes are marvelous seeing devices, there is an important limitation on the resolution of the image in any camera-type high-acuity eye: the phenomenon of diffraction (see Figure 5.3), itself an inevitable consequence of the wave nature of light.

Like an ocean swell entering a harbor, when light waves pass through an opening, or aperture—the pinhole of the *Nautilus,* the pupil of the human eye, or the aperture of a telescope or camera—they suffer dispersion or diffraction and bend outwards after passing through the entrance or aperture.

One consequence, first observed by astronomers, is that when light from a point source (e.g., a star) is focused to an image in a telescope, instead of being focused to a point, it spreads out into a tiny disc, the so-called Airy disc, surrounded by a series of concentric bright and dark rings (see Figure 5.4).[26]

The same is true in the case of our camera-type eye, where the image of a point of light in the field of view forms a disc on the retina. The formation of the disc, whether in the eye or a telescope, reduces the resolving power of the optical device, because, when two point sources in the visual field are close together, their Airy discs may overlap and the two sources cannot be resolved.

It is for this reason that in *all optical devices* that focus an image through an aperture, the resolution of the image is said to be *diffraction limited,* an inevitable consequence of the wave-like nature of light.

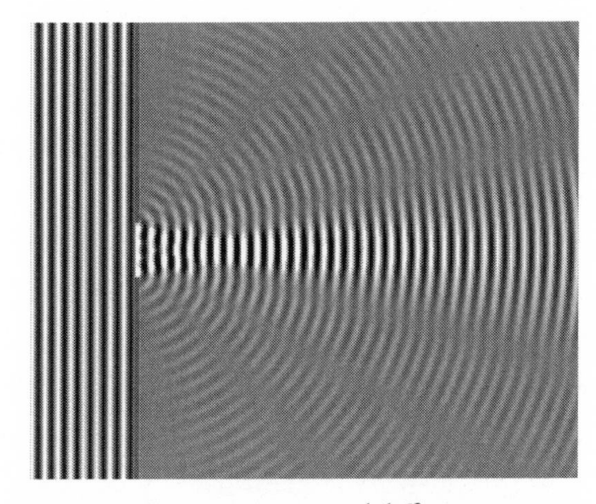

FIGURE 5.3. Computer-generated diffraction pattern.

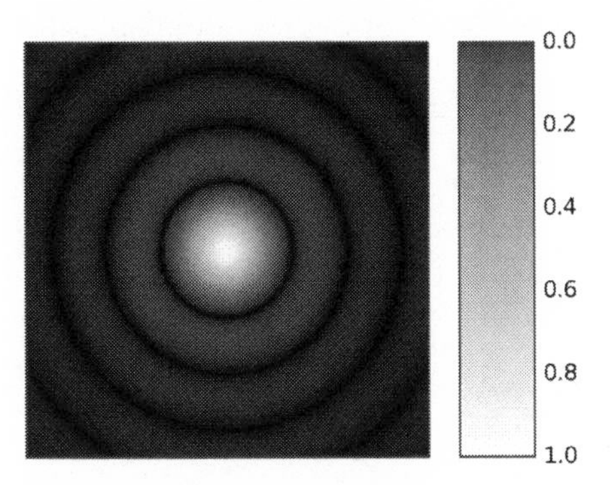

FIGURE 5.4. Image of computer-generated Airy disc. Scale in microns.

The relationship between disc diameter, wavelength, aperture, and focal length is given below in the formula derived from the Airy function.[27]

Airy disc diameter (in microns) = 2.44.λ.F/A
(Where λ is wave length, F is focal length and A is aperture)
Disc diameter (or degree of diffraction) proportional to λ.F/A

From the formula: If F is 1.7 cm (as it is in the human eye[28]) and A ranges from 0.2 to 0.8 cm (the figure usually cited as the range of pupil diameters in man[29]), this gives an F/A ratio ranging from 2.1 to about 8.5. If the wavelength is 0.5 microns (middle of visual band) this gives a range of Airy disc diameters from close to 10 microns to about 2.5 microns.[30] 2.5 microns is the minimum disc size achievable on the human retina and hence corresponds to the maximum resolving power of the human eye. This implies that the minimum size of the Airy disc on the human retina is close to the minimum diameter of the photoreceptors (see above). In most conditions, however, the pupil diameter is close to three millimeters, which gives a disc diameter of close to seven microns.

And just like an ocean swell entering a harbor, where the degree of diffraction of the waves is directly proportional to the wavelength of the swell (the distance between successive waves), in the case of light (and any other type of EM radiation), *the larger the wavelength, the larger the diffractional effect, the larger the Airy disc and the lower the resolution.*

Disc diameter (or degree of diffraction) proportional to λ

And again, as when sea waves enter harbor entrances, the bigger the entrance, the less the waves bend as they enter the harbor, so when light enters the pupil or the aperture of an eye or camera, the smaller the diffractional effect, the smaller the Airy disc, and the greater the resolution of the device.

Disc diameter (or degree of diffraction) proportional to $1/A$

The Problem with Shorter Waves

DESPITE THE diffraction limitation, our visual acuity is still impressive. The maximum resolving power of the healthy normal eye is about one arc minute,[31] equal to one sixtieth of one degree, which is the same as 20/20 vision. Since one degree is 1/360 of a turn (or complete rotation) and one minute of an arc is 1/21600 of a turn,[32] then a disc that is thirty centimeters in diameter subtends one arc minute at one kilometer, and two lines one millimeter apart subtend an angle of about one arc minute at about four meters. The resolving power of the human eye allows us to see readily details on the surface of the moon with the naked eye as the moon subtends thirty arc minutes.[33] And at three meters or about ten feet, most humans with normal vision can easily distinguish between a wasp and a bee. Among animals on Earth, only the raptors have eyes of substantially greater resolving power than man's.

But might we have had eyes of even greater resolving power? Might we have been able to distinguish between a bee and a wasp at thirty meters? Had the wavelength of light been less, then diffraction would also be less and the diameter of the Airy disc proportionately less. This would mean the degree of overlap on the retina would be less, and consequently the resolving power of the eye would be substantially increased.

From the formula—Disc diameter or degree of diffraction proportional to λ—it follows that if the wavelength of light had been ten times shorter, ranging around 0.05 microns, then the minimum Airy disc would be only 0.25 microns (250 nanometers) across, and if 100 times shorter, around 0.005 microns, the Airy disc would be only 0.025 microns (25 nanometers).

But unless the diameters of the photoreceptors in such a counterfactual world were reduced to the same degree, to diameters of 0.2 microns and 0.02 microns respectively—i.e., unless the pixel size was correspondingly less—there would be no gain in resolution.

But reducing the diameter of photoreceptors in an attempt to improve resolution would be problematic, because any substantial reduction in the diameter of the tubes would translate into a massive decrease in their volume (by a factor equal to the decrease in the linear dimension squared). *All other things being equal,* reducing the diameter of a human rod receptor (two microns in diameter) by ten times to 200 nanometers would necessitate a corresponding reduction in the volume of the cell by 100. Then, instead of there being room for 150,000,000 light-absorbing molecules (rhodopsin) in the rod outer segments of the receptors there would be room for only 1,500,000. If the diameter were reduced 100 times to a diameter of 20 nanometers, the reduction in volume would be 10,000 times and a photoreceptor in this extreme counterfactual Lilliputian world[34] would only have room for 1,500 molecules! The fundamental and intractable problem in any size reduction is that the size of the molecules, cell membranes, and molecular machinery of photoreception in canonical carbon-based cells on Earth cannot be reduced by a linear factor of ten, let alone a factor of 100.

The atoms will not oblige. They are *exactly the same size* in a red blood cell as in a grain of dust in the Andromeda Galaxy.

Carrying out equivalent functions in photoreceptors of greatly reduced volume, and hence greatly reduced number of photon-capturing molecules, would clearly be very challenging! The photoreceptors are among the most complex of cells in the body and the retina one of the most complex of organs, involved not just in transmitting visual information to the brain but in sophisticated analysis and modulation of the image. Any reduction in photoreceptor volume and in number of receptor molecules in the detector tube, necessitated by a reduction in cell diameter, would greatly reduce our ability to see over the vast range of luminosity experienced in nature—a range of 10^{12}.[35] We would lose our ability to see an individual photon at night with our rods and capture a million photons a second to see in the brightest conditions with our cones.

Moreover, engineering tubes much narrower than mammalian photoreceptors might pose serious problems. The only known long, narrow, bio-tubular forms much smaller than photoreceptors in present-day life forms are cilia and flagella, about 250 nanometers across[36] (or about one quarter the diameter of the narrowest sections of the photoreceptors in the human eye, the outer segments of foveal cones[37]), and microtubules, which are twenty-five nanometers across and have an internal diameter of twelve nanometers (about one-tenth the diameter of a cilium). But in no known microtubule is any physiological function carried out *inside* the tube. This is not surprising, as, for example, the bilayer cell membrane is ten nanometers across and a ribosome thirty nanometers across! Attempting to engineer functional photoreceptors—tiny optic fibers—remotely as small in diameter as a microtubule would almost certainly pose insurmountable bio-engineering problems! The possibility of reducing a rod photoreceptor diameter 100 times to twenty nanometers and preserving any sort of light-detecting bio-function remotely equivalent to that of a normal sized photoreceptor can be completely discounted.

This *gedanken* (thought) experiment implies that in a counterfactual world, where light waves were ten or more times shorter, even though the *optical* resolution of the image on the retina would be several times better, there would be no improvement in visual acuity of the human eye or any high-acuity biological eye, because of the fundamental constraints related to the minimum size of functioning bio-components and the inescapable relationship between the linear dimension and the volume of the photoreceptor cell. Simply put, the pixel size in the retina can't be reduced significantly.

It is certainly an arresting fact that the minimum size of the Airy disc in our eyes and those of other vertebrates (generated by the diffraction of visual light of wavelength ranging round 0.5 microns passing through pupillary apertures of a few millimeters) corresponds so closely with what is probably nearly the minimum diameter of biological detector cells.

The Problem with Longer Waves

ENVISAGING HIGH-ACUITY eyes in biological beings of approximately our dimensions if the wavelength of visual light were longer is just as problematic!

If the wavelength of light, instead of ranging round 0.5 microns, were around five microns (ten times longer) or fifty microns (100 times longer), the resolution of the eye would be very much less because of diffraction and the diameter of the Airy disc would be 10–100 times greater (25–250 microns across respectively in the human eye), as discussed earlier. Instead of one Airy disc covering about one or two photoreceptors in the retina, each disc would cover approximately 100 receptors and at 100 times longer, 10,000. In short, the acuity of the eye would be very much diminished!

To compensate for the increased diffraction and lower resolution and recover the resolving power of the eye, *only one strategy* is available: to increase the aperture of the eye by ten times (for light of wavelength 10 times longer, around 5 microns) or 100 times (for light of wavelength 100 times longer, around 50 microns).

Disc diameter or degree of diffraction proportional to λ/A

But this would necessitate, on a tenfold increase in wavelength, an eye ten times larger—the size of the human head and, on a hundredfold increase, one hundred times the diameter, the size of a small car! The very largest eye of any organism on Earth is that of the giant squid, which is twenty-seven centimeters in diameter or just over ten times the diameter of a human eye. The second-largest eye is that of the blue whale, which is eleven centimeters in diameter. The largest eye of any fish is that of the swordfish, at nine centimeters.[38] The largest eyes of terrestrial organisms are much smaller, four centimeters in a moose, and 3.5 centimeters in an elephant or horse.[39]

Moreover, an increase of ten and certainly 100 times in the diameter of our eyes would be incompatible with our upright android design and basic dimensions. On a tenfold increase, to conserve the proportion between eye size and head size in an upright android being, we would have to be as big as the giant inhabitants of Brobdingnag (the island of giants) in Swift's *Gulliver's Travels,* which were about twelve times the height of Gulliver.[40]

But such giant humans are a fantasy. With Earth's current gravity, elephants need massive legs to support their great mass. At about twelve times our height, Swift's giants would weigh 1,000 times more, close to 100,000 kilograms—ten times the weight of an African elephant. But in addition to the non-feasibility of massively increased limbs, there is another reason why giants are impossible (as was discussed in Chapter 4 of *Fire-Maker*).

Tripping would be a catastrophe for such giants, as their massive heads (1,000 times the volume of a human head and weighing in the order of 1,000 kilograms) would hit the ground with such force that the skull would be fragmented and the brain disintegrate. As Fritz Went pointed out in an *American Scientist* article some time ago, the increased danger of serious injury by tripping limits the height of any upright android form:

> A 2 m tall man, when tripping, will have a kinetic energy upon hitting the ground 20–100 times greater than a small child who learns to walk. This explains why it is safe for a child to learn to walk; whereas adults occasionally break a bone when tripping, children never do. If a man were twice as tall as he is now, his kinetic energy in falling would be so great (32 times more than at normal size) that it would not be safe for him to walk upright.[41]

Even as things stand, many serious skull injuries result when people trip and hit their head on the ground. Steven Vogel makes the same point in his *Comparative Biomechanics,*[42] noting that tripping is a serious hazard even for cows and horses, animal much smaller and lighter than Swift's giants.

As pointed out in *Fire-Maker,* Chapter 2, only rocky planets of approximately the size of the Earth, with the same gravitational pull, can hold an atmosphere consisting of heavy gases like oxygen and nitrogen and support advanced, active carbon-based life forms like ourselves. If they are bigger, they retain the primeval gases like Jupiter. If they are smaller, they lose their atmosphere like the Moon. No tiny planet with a gravity much less than that of the Earth that might facilitate the existence of giants and actualize Swift's fantasy would have the right atmosphere. The same gravitational force which retains the vital oxygen for life also imposes an absolute limit on the size of terrestrial organisms.

Clearly, given the constraints imposed by gravity on the size of android-type beings on habitable planets, if the wavelength of light with the right energy levels for life had been longer by a factor of more than ten, then a high-acuity eye with a resolving power close to that of the current human eye would have been simply too big for any android being that might inhabit an Earth-like planet of the right size to possess an oxygen-rich atmosphere.

But in addition to the difficulty of attempting to recover high resolution by massively increasing eye size and lens diameter, there are many different types of optical aberration, familiar mainly to optometrists and ophthalmologists (astigmatism, coma, etc.), which decrease image quality. These arise from minor defects in the shape and configuration of biological lenses—consisting of globular masses of biological matter (i.e., wet ware)—which are not present in man-made optical systems.[43] In the case of the human eye, these are minimal when the pupil diameter is small, but are enhanced as aperture increases.[44] Given the inevitable minor distortions in any biological lens, it is not hard to imagine just how severe these would be in a biological lens more than ten times bigger.

Again, the necessity of a lens, which is made up of living tissue and which must obtain its nutrients by diffusion, must also place an upper limit on the size of a vertebrate camera-type eye in any biological system.

Then there are the additional biological constraints in maintaining the basic round shape of the eye itself. As I pointed out in *Nature's Destiny*, the maintenance of the globular form of the eye in many vertebrates, "is partly achieved... by placing the eye in a round bony socket (which also preserves the eye's delicate structure from damage)."[45] In many vertebrate species with relatively large eyes, including whales and elephants, the sclera (the outer membrane surrounding the vertebrate eye) is greatly thickened to conserve the form of the eye.[46] And as I further pointed out: "Another adaptation which helps maintain the spherical shape of the eye is the maintenance of a relatively high hydrostatic pressure in the interior of the eye."[47] In this context it is also worth noting that even the shape of a glass refracting lens bigger than a meter will distort under its own weight.[48]

Summary Regarding Wavelength

It would seem, then, that if the wavelength of light were more than ten times longer, compensating for the inevitable loss in resolution by increasing the dimensions of the eye, and more specifically the diameter of the lens, would be impossible due to basic biological constraints, while if the wavelength had been ten times shorter, no gain in resolution would be possible because of another suite of biological constraints imposed by the minimum size of *biological* molecules and cellular structures.

In short, given the basic constraints of biology, the wavelength of light is almost exactly what it needs to be for high-acuity vision in organisms of our approximate size and biological design, inhabiting a planet of the right size and gravity to maintain an oxygen-rich atmosphere capable of sustaining advanced carbon-based life.

The Right Light—Again

Given the immense range of wavelengths in the EM spectrum, from a fraction of a nanometer in the gamma region to tens of kilometers in the radio regions, even if we allow the possibility of high-acuity vision in beings of our biology and size utilizing different sets of wavelengths anywhere between 0.05 microns to five microns, this extended counter-factual range would still be *an infinitesimal fraction of the entire stretch of the EM spectrum.*

These considerations have led us back to the same tiny magic band that has just the right energy levels for photochemistry and detection by bio-matter. We have yet again had to select the same playing card from the stack that stretches that inconceivable distance beyond our nearest neighboring galaxy.

And this is a genuine coincidence in the nature of things. The laws of nature that determine that EM radiation in the visible spectrum should have just the right energy levels for photodetection are quite different from the laws of optics that determine that only light of wavelengths between 0.380 and 0.760 microns[49] is fit to form an image in high-acuity eyes of size commensurate with beings of our size. And the laws of nature that determine that rocky planets the size of Earth should be fit to retain an atmosphere and allow the standing tall of an android form of our weight and dimension represent another entirely independent set of laws.

Further Constraints. The energy levels and wavelength of light are of course only two aspects of the natural order that must possess precisely the values they do for high-resolution vision. There is also the transparency of water to visual light, a factor critical to photosynthesis as well as vision. There is the low refractive index of water, which is the major constituent of every biological lens and confers upon the lens its capability to focus an image. If the refractive indices of water and bio-matter had been less or if light had travelled at the *same speed* through water as through air, there would be no refraction and no lens and no retinal image to process and no high-acuity vision. Again, the lens itself is a living tissue but has no blood vessels (as these would greatly degrade the image), and hence relies on the *right diffusion rates* of oxygen and small organic molecules in bio-matter. A large, sophisticated nervous system to analyze the visual data is also necessary, which is in turn dependent on the high metabolic rates of air-breathing organisms, which is again only possible because of the oxygen provided by photosynthesis, i.e., by the light!

Quantum Fitness for Vision

As MENTIONED in Chapter 2, biologists for most of the past century have been able to account for the vast majority of biological phenomena at the molecular level in terms of classical particles of matter, without any recourse to wave-particle duality or quantum weirdness.

The one obvious exception to this has been vision. The focusing of the image on the retina, the formation of the Airy disc, and the refraction of light through a lens have long been explained classically in terms of light "waves." On the other hand, light's particulate *alter ego* explains how light can be detected or captured as individual packets of energy by the molecular receptors in individual photoreceptors situated at particular points on the retina. The discovery by George Wald[50] in the 1930s that individual light photons are detected by individual retinal pigment molecules in the photoreceptor cells was one of the landmark advances in visual physiology.

Because you can only explain the formation of an image by assuming that light behaves like a classical wave and only explain its detection by assuming it behaves like a classical particle, in all biology texts published over the past several decades there has been an implicit (if unacknowledged) acceptance of this mysterious duality in accounting for the phenomenon of high-acuity vision.

There is a profound difficulty in imagining how any phenomenon remotely comparable to vision, which provides the most comprehensive and detailed picture of our environment (more than that of any other sense), could be possible if light did not exhibit the two contradictory personas. Without quantum weirdness, nature would not be fit for high-acuity vision; we would lack the gift of sight. And as high-acuity vision is of utility *only* for complex, advanced forms of life, wave-particle duality can be seen to be yet another element of fitness in the order of things for complex beings of our biological design. The "rock eating" denizens of the dark (see Appendix B) who get by perfectly well without sunlight have no need to see in that eternal night of the ocean depths.

This is also relevant to photosynthesis, which provides the oxygen we need to sustain the very high metabolic rates necessary for vertebrate vision (see below). The *particulate persona* of light is manifest when individual photons are absorbed by the chlorophyll molecules (see the previous chapter), while the *wavelike persona* of the electron is manifest in their quantum tunneling to transit from the chlorophyll molecules to the reaction centers in the leaf. (See the discussion in Chapter 2.)

One final, fascinating point: Our vision is not only dependent on the wave-particle duality of matter. It is also dependent on the fact that the EM wavelengths in the visual region that are fit for making an image in an optical device of the size of the human eye have just the right energy levels for biological detection. This depends in turn on the Planck constant, one of the grand constants of physics, having *just the value* it does—the unimaginably small value close to 10^{-35} joule-seconds.[51] It seems the physical parameters that enable human vision were built into the basic fabric of things from the moment of the Big Bang.

Summary

On any consideration of the fitness of nature for high-acuity vision, it is impossible to escape the conclusion that nature does indeed exhibit a special fitness for beings of our physiological and anatomical design. The facts convey a powerful impression that the basic order of things was designed to that end. A skeptic might point out that biological lenses are imperfect and that the resolving power of the eye is not infinite. But no lens—biological or otherwise—can be *perfect* and no eye can possess *infinite resolving power.* Indeed, *no optical device*—biological or otherwise—can have *infinite resolving power.* To distinguish a bee from a wasp at one kilometer would require an eye many times larger than the human eye, and to see individual stars in the furthest galaxies, 13,000,000,000 light years away, would necessitate an eye far bigger than any current radio telescope.

But despite their limited powers and the inevitable imperfections of any biological lens, our eyes have served us well. They have led us from the time of our Stone-Age, tool-making ancestors through the conquest of fire and the development of technology and science to the twenty-first century. And what cannot be denied is that the possession, by biological beings of our android design and dimensions, of a high-acuity eye (even if not capable of seeing to the end of the universe) depends on an extraordinary fortuity in the order of things.

As shown in Figure 5.5, the minuscule band of EM radiation, *representing a one in 10^{24} fraction of the EM spectrum,* is exactly right for photosynthesis, the giver of the vital oxygen for advanced, carbon-based life (and specifically for the functioning of the light-hungry photoreceptors), and is also exactly right for a high-acuity optical device in a biological organism of our size and android design (for completely different reasons). This is the prime coincidence which launched the human enterprise. Take away oxygen and no advanced, intelligent, carbon-based organisms depending on oxidation to supply their extravagant metabolic rates would be possible. Take away vision and no species capable of taming fire and developing an advanced technology would be possible.

Our uniquely human desire for knowledge could only have been fulfilled by the gift of sight. Virtually all our knowledge of the world, and particularly our scientific knowledge acquired over the past four centuries, has been largely dependent on our possession of eyes of high resolving power and capable of bringing us a very detailed and information-rich image of our surroundings.

Seeing is expensive, however. The oxygen consumption of the mammalian retina (per gram of tissue) is nearly fifty percent greater than that of the kidney, nearly three times greater than that of the cerebral cortex, and nearly six times that of cardiac muscle.[52] G. L. Walls describes the photoreceptors in his classic *The Vertebrate Eye* as "greedy,"[53] and greedy they are, for both nutrients and oxygen. Indeed, the high acuity and high sensitivity of the visual system in higher vertebrates is critically dependent on the very high metabolic rate or oxygen consumption of the photoreceptors.

It is surely deeply satisfying and wonderfully parsimonious that light enables vision in two ways: First, light provides via photosynthesis the vital oxygen necessary to generate the very high metabolic rates which enable photoreception in the human eye; and second, the very same light forms the image on the retina and gifts us with the miracle of high-acuity vision.

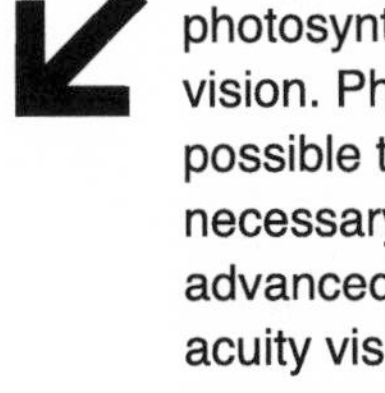

FIGURE 5.5. Visual region for photosynthesis and high-acuity vision.

Figure 6.1. The Ancient of Days, by William Blake, originally published in 1794.

6. The Anthropocentric Thesis

In a universe whose size is beyond human imagining, where our world floats like a dust mote in the void of night, men have grown inconceivably lonely. We scan the time scale and the mechanisms of life itself for portents and signs of the invisible. As the only thinking mammals on the planet—perhaps the only thinking animals in the entire sidereal universe—the burden of consciousness has grown heavy upon us. We watch the stars, but the signs are uncertain. We uncover the bones of the past and seek for our origins. There is a path there, but it appears to wander. The vagaries of the road may have a meaning, however; it is thus we torture ourselves.

—Loren Eiseley (1946), *The Immense Journey*[1]

This book has documented some of the remarkable ways in which nature is fit for life on Earth and particularly for beings of our biological design. Although the focus has been on the fitness of light for photosynthesis and high-acuity vision, there are many additional elements of fitness in nature for our type of being, many equally remarkable, some of which I have reviewed in previous publications. But despite the evidence, the *anthropocentric claim* that we occupy a special place in nature is currently very unfashionable, among both scientists and the general public.

Nothing symbolizes more graphically the demise of the anthropocentric belief that modern humans occupy a special place in the world order or that nature is uniquely fit for beings of our biology than modern science fiction. No one who has seen the first *Star Wars* movie will ever forget the bizarre zoo of alien bio-forms drinking alongside Han Solo in the Cantina bar. Or who can forget the extra-terrestrial hiding in the closet in a suburban L. A. home in Spielberg's masterpiece *ET*? Even more strange are the myriad of non-anthropoid aliens in *Star Trek* that live in toxic environments, exist as pure energy without physical form, or are so foreign they are not recognized as alive at all when first encountered.

The denial that there is any special fitness in nature for life on Earth, or for beings of our biology uniquely possessed of consciousness and intelligent agency, not only sanctions the notion of alien life forms, but also sanctions strong AI and the notion of machine intelligence[2] and sentience and the associated fear of a machine takeover raised by Nick Bostrom in his recent book *Superintelligence*.[3] These also make appearances in popular media, as characters like *Star Wars'* C-3PO, Hal the conscious supercomputer in *2001*, Sonny in *iRobot*, or Data in *Star Trek: The Next Generation*.

Such scenarios are, of course, pure science fiction, but the underlying notion that the cosmos is fit for a vast zoo of alien life-forms of wildly differing biologies and biochemistries as well as intelligent mechanical forms, is not science fiction but a world view that suggests *that there is no special fitness in nature for intelligent, conscious agents like ourselves*.

According to the more extreme versions of this conception, the cosmos teems with complex alien life-forms, some based on exotic chemistries and some immensely more intelligent than ourselves. We envisage them as the creators of technological civilizations millennia ahead of our own, whose machines would seem in comparison to our own creations as television would seem to a Neanderthal, their technology like a type of magic.[4] The underlying belief is that terrestrial intelligent life and its supreme manifestation in mankind is merely "*one voice in the Cosmic Fugue*," as Carl Sagan put it so eloquently.[5]

Despite the delightful, unrestrained speculation of science fiction movies (and some only slightly more restrained ideas in some recent books[6]) there is a very widespread scholarly consensus among biologists and biochemists that carbon is perhaps the only fit atom for building organic chemicals, and that organic chemicals are uniquely fit to form the molecular building blocks for highly complex chemical replicating systems. This consensus has existed for at least one century, since Lawrence Henderson and Alfred Russel Wallace.[7] A long list of first-class scholars[8] following Henderson and Wallace have also defended what Carl Sagan called "carbon chauvinism," which Sagan conceded was his own view.[9] Among physicists and cosmologists, there is also a scholarly consensus that the laws of nature are fine-tuned for life, at least generic chemical life, and for the formation of rocky planets like Earth that provide the right planetary conditions for chemical life to thrive, including but not necessarily restricted to the form of carbon-based life as it exists on Earth.[10]

The physical evidence alluded to by the physicists is certainly very impressive and endlessly fascinating, revealing that unless the laws of nature were exactly as they are, the universe would contain no stable solar systems, no rocky planets like the Earth, no atoms, molecules, or chemistry, and no forms of life that we can imagine. But in itself the evidence from physics alone can never answer Thomas Huxley's celebrated "Question of questions... the ascertainment of the place which Man occupies in nature and of his relations to the universe of things."[11]

Unfortunately, the biological evidence which might answer Huxley's question is seldom considered by biologists. The great majority are averse to the notion of fitness. This is undoubtedly due to the pervasive influence of Darwinism within the biological community and the notion that all design in the biological realm is the result of adaptation to environmental conditions via natural selection, a notion which is the very converse of the conception that the natural environment is uniquely fit (designed) for life on Earth and for beings of our biology and was so long before life originated or natural selection could operate.

One encounters this aversion—to any hint that the natural order might be uniquely fit for life or mankind—throughout biological literature. One recent example that came to my notice, which seemed extraordinarily incongruous, was in a paper arguing for the "inevitability of complex, macroscopic [carbon-based] life"[12] which omitted any serious discussion as to whether the series of "inevitable" events it described might depend on any special facilitation or fitness in the order of things.[13] Land plants, for example, which the authors describe as a major innovation in the history of life on Earth, depend on the fitness of not just one but a whole suite of unique properties of water (see *The Wonder of Water,* Chapters 1 and 5). Take away any one of these elements of fitness and plants would never have emerged from the oceans. But these facts are completely passed over. Again, in the wonderful account of the biophysics of the cell given in *Physical Biology* by Phillips et al.,[14] although the phenomena alluded to on nearly every page depend on a fantastic fortuity in the order of things, there is no mention of fitness. There is no mention, for example, of the fact that the functioning of the cell's molecular machinery is dependent on the precise relative strength of the weak and strong chemical bonds being almost exactly as they are.

Because biologists are so averse to the notion of fitness and because the evidence of physics by itself cannot secure a significant place for beings of our biology in the universe, Huxley's question has gone unanswered.

It is certainly curious that, despite the almost universal acknowledgement among physicists and cosmologists that the cosmos is uniquely fine-tuned for generic chemical life and the widespread acknowledgement among biochemists that carbon is uniquely endowed with the right properties for chemical replicating systems, there is a complete failure to take the next logical step. The possibility that nature may be also fine-tuned specifically for advanced forms of carbon-based life like ourselves is almost entirely ignored. Indeed, I would argue, as I did in *Evolution: Still a Theory in Crisis*[15] that this failure is *one of the most striking failures of the human imagination in recent scientific history.*

And it is precisely because no one is prepared to take the next step and consider the evidence of nature's special fitness for our type of advanced intelligent being that the widespread illusion persists that intelligent aliens could be nonhuman and perhaps hugely different from ourselves. This is evidenced by the fact that the designers of Pioneer 10, the

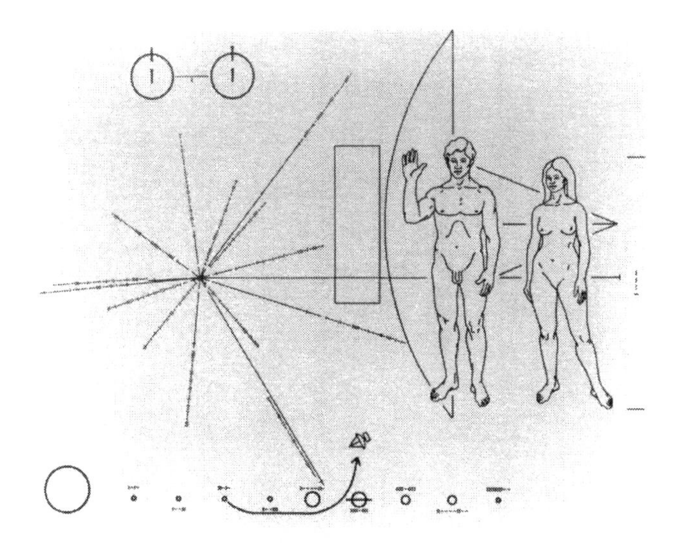

FIGURE 6.2. One voice in the cosmic fugue? Pioneer 10, launched in 1972, was the first space craft built by mankind destined to leave the solar system. Its designers, believing there was nothing unique or special about our biology and that the laws of nature are compatible with many other types of life and intelligent beings, fixed a plaque (shown above) on the space craft informing any aliens of our own particular biological design. The same mindset led to the encrypted digital message sent out from the Arecibo observatory in 1974, indicating our biology and the structure of DNA (see Figure 6.3). The Pioneer 10 plaque was designed by Carl Sagan and Frank Drake.

first machine built by mankind destined to leave the solar system, felt it necessary to fix a plaque on the space craft informing the aliens out there of our own peculiar biology (Figure 6.2). Precisely the same *Zeitgeist* led to the encrypted digital message sent out from the Arecibo observatory indicating our biology and the structure of DNA (Figure 6.3).

The graphic images of alternative life depicted in science fiction films, along with the more serious speculations of exo-biologists, futurists, and advocates of AI, reveal just how far Western culture has travelled in the four and a half centuries since that fateful year, 1543, when Copernicus published *De revolutionibus*. Our view of ourselves has shifted, from our

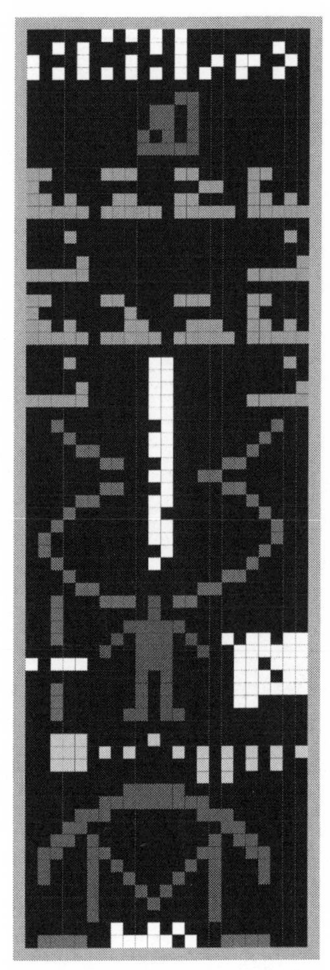

FIGURE 6.3. Digital message sent out by the Arecibo radio telescope.

being the central focus of the celestial dance to an epiphenomenal, freakish, and hugely improbable product of chance and time; from the image of God, a necessary being preordained from the beginning, to a contingent entity no more significant in the cosmic order than his own mechanical creations or a transient, wind-blown pattern of autumn leaves. We see ourselves as an artifact in a universe which knew nothing of our coming and cares nothing for our fate.

This current secular vision, which might be considered one of the defining ideas, if not *the* defining idea, of our time—the *Zeitgeist* of our twenty-first century civilization—is as far removed from the anthropocentric cosmos of the medieval scholastic philosophers—the notion of a world order specially configured for our existence—as could be imagined, and surely one of the most dramatic intellectual transformations in the history of human thought.

The Anthropocentric Thesis

No matter how powerful the grip of the post-Copernican, secular demotion of man to a cosmic afterthought on the popular imagination, and no matter how great its appeal in secular quarters, the facts of science provide a massive testimony against it. No matter how unfashionable the notion may be in many intellectual circles, the evidence is unequivocal: Ours is a cosmos in which the laws of nature appear to be specially fine-tuned for our type of life—for advanced, carbon-based "light eaters" who possess the technologically enabling miracle of sight!

I do admit that the claim—that our existence depends on a profound fitness in nature for *our specific form of being*—is among the most outrageously ambitious claims in the history of thought. Could the cosmic dance have really been arranged primarily for beings like us? The claim is indeed *extraordinary* and, as Carl Sagan rightly argued, "Extraordinary claims require extraordinary evidence."[16]

Showing that there is, indeed, extraordinary evidence has been the main aim of this book and the preceding books, *Fire-Maker* and *The Wonder of Water*. In *Fire-Maker*, we saw that the size and atmosphere of our planet, the fitness of Earth for large trees, and our own physical design all prepared the way for our technological journey from the Stone Age to the advanced civilization of the twenty-first century. And in *The Wonder of Water*, we saw how water's diverse and often anomalous properties serve a myriad of crucial roles in the development and sustenance of life on Earth, including some very specific properties which make possible our own physiological design.

In this book, I have described the fitness of the radiation emitted by the Sun for life on Earth: the fitness of the atmosphere to ensure sufficient IR radiation is absorbed to warm the surface of the planet, preventing water from freezing and animating matter for chemical reactions; the fitness of the atmosphere to let through the visible light to the Earth's surface to enable photosynthesis which generates the oxygen and reduced carbon fuel necessary to support our "light eating," energy-demanding lifestyle; and the fitness of the same light for high-acuity vision in beings of our bodily design and size.

I believe that any discerning reader who carefully considers the evidence presented here and in *Fire-Maker* and *The Wonder of Water* will concede that no matter how unfashionable the idea, no matter how extraordinary the claim, the facts speak for themselves. No matter how incongruous in the context of the current secular *Zeitgeist*, the cosmos as revealed by modern science is a cosmos which is extraordinarily fit in so many ways for beings of our anatomical and physiological design.

It is important to note that the claim that nature exhibits an extraordinary fitness for *our type of advanced, intelligent, carbon-based life forms* does not exclude entirely the possibility of other types of primitive chemical life based on, say, silicon (or indeed some forms of simple mechanical life). My only claim is that on the evidence available, nature would appear to exhibit a very special fitness in so many diverse ways for beings of our physiological and anatomical design.

Moreover, there must surely be many more elements of fitness for our biology which await discovery. This expectation is borne out by the fact that *every major advance in science since the beginning of the nineteenth century* has revealed further elements of unique fitness in nature for life on Earth. The founding of modern chemistry led in the early nineteenth century to the discovery of the uniqueness of the carbon atom. In the late nineteenth century and beginning of the twentieth century, further advances led to the discovery of the unique fitness of carbon biochemicals in a water medium for the assembly and functioning of complex chemical systems—a fitness described in A. R. Wallace's *World of Life* and Lawrence Henderson's classic *The Fitness of the Environment*. During the twentieth century, advances in astrophysics and cosmology led to the discovery of the fine-tuning of the basic constants of physics for a universe capable of sustaining life. The elucidation in the mid-twentieth century of the mechanism of nuclear fusion in stellar interiors revealed that the synthesis of carbon and the higher atoms of life depended on a special fitness to that end in the nuclear energy levels of the key atoms involved. Then there was the discovery of plate tectonics, again in the mid-twentieth century, which revealed the underlying mechanism by which the chemical constancy of the Earth's hydrosphere is maintained, another vital element of fitness for the habitability of rocky planets like the Earth. Advances in the next few decades are sure to reveal additional unique elements of fitness, perhaps involving the strange quantum behavior of matter at the subatomic level.[17]

But whatever future discoveries might bring, I rest my case. The evidence is already compelling: Whatever the cause and whatever the ultimate explanation, nature appears to be fine-tuned to an astonishing degree for beings of our biology. The ground of being, it seems, was fabricated for our existence from the moment of creation. The universe, in the words of Freeman Dyson, must have known "we were coming,"[18] not just generic carbon-based life forms, but beings of our physiological and anatomical design! On any consideration of the evidence, science has revealed a universe that echoes closely the anthropocentric world view of the medieval scholars, who saw the universe as an intelligently created natural unity in which man was no epiphenomenal, contingent product of deep time and chance, as he is widely considered today, but rather its very end and purpose.

Appendix A. Doing Without Sunlight

There was no more dramatic reminder that exotic life (from our perspective) can thrive in the dark than when, in 1977, two geologists aboard the deep-sea research submarine *Alvin,* exploring the Galapagos Submarine Ridge in the East Pacific, made a discovery, sensational at the time. They found that organisms of many types, constituting whole complex ecosystems, *thrive and maintain themselves in the dark without any direct solar energy input* in the ocean depths.

The geologists onboard *Alvin,* John Corliss and John Edmond, were not looking for exotic life forms. They were on a geological mission, searching for evidence that sea water is being continually circulated through hot volcanic rocks (as this was seen to be one explanation of the very different composition of sea and river water), and were, as described by Gross and Plaxco, naturally very excited to find that the sea temperature outside their submarine, close to the Galapagos submarine ridges, was anomalously high.

Gross and Plaxco describe what happened next:

> This hint of hot springs on the ocean floor was thus a sensational discovery, and the researchers excitedly took samples so that they could later determine the chemical composition of this unexpectedly warm water. Still excited, they piloted *Alvin* up to the top of the ridge, where a much bigger sensation was waiting for them. Where they had expected to find a stark "desert" of bare, lifeless basalt, freshly erupted from the spreading center atop the ridge, they found instead an oasis 100 meters in diameter, with warm water sifting

through every tiny crack in the seafloor, and richly populated with clams, crabs, sea anemones, and large pink fish."[1]

When they excitedly reported their finding back to Woods Hole Oceanographic Institution, the marine biologists there first refused to believe their story. The pilots in *Alvin* were geologists, after all![2] But what they had seen was no illusory vision; it was the first glimpse of a complex alien biota independent of the energy of the Sun.

Before this discovery, it was assumed that life in the deep ocean floor was dependent on what is termed "marine snow"—organic matter and detritus which rains down from the sunlit regions of the upper ocean. Organisms feeding on this detritus would still be dependent on photosynthesis and the light of the Sun. However, the flux of material due to marine snow is quite insufficient to provide nutrients for anything but a tiny fraction of the biomass of organisms which the researchers on *Alvin* had seen clustered around the hydrothermal vent.[3]

Subsequent research revealed that the complex metazoan life forms (such as clams, crabs, and fish) which live in the surrounds of the vents survive by feeding on organic compounds synthesized by a zoo of exotic microbial organisms endemic to the vents. These derive their nutrients and energy from inorganic compounds bubbling up in the hot waters of the vents.[4] These exotic bacteria are known as chemosynthetic or lithotrophic life forms, and although many do use molecular oxygen derived ultimately from photosynthesis, many, as Norman Pace points out in a *Science* review article, derive energy by oxidizing reduced inorganic substances such as hydrogen, hydrogen sulphide, or ferrous iron, utilizing nitrate, sulphate, sulphur, or carbon dioxide instead of molecular oxygen as the terminal electron acceptor.[5] Because the reduced foodstuffs are leached initially from the crustal rocks underlying the vents, the lifestyle of these bacteria is termed lithotrophy or "rock eating." Such "rock eating" anaerobic bacteria are *completely independent of the Sun*. As Minic and Hervé comment further:

> The aptitude of living organisms to survive and constitute an important biomass around hydrothermal vents is linked to the unique

> chemistry of these environments. Sea water penetrates into the fissures of the volcanic bed and interacts with the hot, newly formed rock in the volcanic crust. This heated sea water (350–450°C) dissolves large amounts of minerals. The resulting acidic solution, containing metals (Fe, Mn, Zn, Cu... [ellipsis in original]) and large amounts of reduced sulfur compounds such as sulfides and H_2S, percolates up to the sea floor where it mixes with the cold surrounding ocean water (4°C) forming mineral deposits and different types of vents. In the resulting temperature gradient, these minerals provide a source of energy and nutrients to chemoautotrophic organisms which are, thus, able to live in these extreme conditions.[6]

Many species of bacteria grow into thick mats near the hydrothermal vents, which attracts other organisms, such as amphipods and copepods, which graze upon them directly.[7]

Of all the metazoan denizens of this sunless realm witnessed by Corliss and Edmond, none were more the stuff of science fiction than the giant tube worms. Belonging to the phylum Annelida (to which familiar earthworms also belong), they are giant parasites growing up to two meters long, having no mouth or digestive system, and obtaining all their nutrients from bacteria which live in their tissues. The bacteria living inside the worm—in an organ known as the trophosome—oxidize hydrogen sulphide (H_2S), derived from the water bubbling up from the vents, to sulphate (SO_4) as a means of obtaining metabolic energy, which they then utilize to synthesize organic compounds that are passed to their hosts.[8]

Many microbial species had already been described before 1977 that use inorganic nutrients rather than reduced carbon compounds as an energy source, as well as many anaerobic species which use terminal electron acceptors (oxidants) other than oxygen. Sulphate-reducing bacteria were first described in the nineteenth century[9] and anaerobic, methane-producing bacteria (methanogens) had also been described decades before 1977. But no one had ever imagined there could be such a rich and complex ecosystem dependent on energy provided by "rock eating."

The paradigm that all life on Earth lives by the sunlight had been overturned. The shock was considerable. As recently as 1959, less than twenty years before *Alvin*, the conventional view was, as formulated by Nobel Laureate George Wald:

> [W]ithout the possibility of photosynthesis how could [living organisms]... ever become independent... and fend for themselves? Inevitably they must consume the organic molecules about them and with that life must come to an end.
>
> It may form an interesting intellectual exercise to imagine ways in which life might arise, and having arisen might maintain itself, on a dark planet; but I doubt very much that this has ever happened, or that it can happen.[10]

How dated Wald's remarks now seem! Life as manifest in the bacterial biota thriving around the submarine vents, albeit of a simple kind far less complex than our own, has thrived for billions of years in the dark of the abyssal depths. So, it is now possible, with apologies to Wald, to imagine a dark world in which carbon-based life forms, albeit simple and unicellular, thrive, forms which, if Nick Lane is to believed, may have originated in the eternal darkness of the hydrothermal vents.[11]

However, there is one important caveat. Unlike the bacterial denizens of the dark on which they feed, *all* the metazoans (the tube worms and fish and crabs etc.) which live round the vents are aerobic and *still depend on oxidation* to generate their metabolic energy (ATP), and *all* use free O_2 as the electron acceptor, and their oxygen is ultimately derived from photosynthesis in the surface water kilometers above the sea bed or in the leaves of terrestrial plants. And as mentioned above, the same is true of many of the microbes, even if the nutrients they oxidize are inorganic. As Charles Smith comments:

> While deep-sea communities have often been characterized in the literature as "life without the sun," the notion is only partially true and can be misleading. Almost all of the observable biomass associated with deep-sea communities such as hydrothermal vents, cold seeps, whale falls and wood falls depends on the action of aerobic chemosynthetic organisms which could not survive in complete iso-

lation from the sun due to their oxygen requirement.... However, even though the vast majority of life in these communities would die out without the sun, I believe that the more significant result of the discovery of these communities remains intact. To focus on aerobic chemosynthesis in the deep-sea is predictable, given how dominant the process is, but ignores the fact it is not the only chemosynthetic lifestyle found at such sites. Anaerobic chemoautotrophs obtain both their electron acceptors and donors from geothermal processes, severing the last link to solar energy that their aerobic cousins could not. What is truly significant about deep-sea chemosynthetic communities is not "how much" life lives in isolation from the sun, but the fact that *any life at all accomplishes that feat.* Although these small and seemingly insignificant microorganisms do not have the same impact in the deep as their aerobic counterparts, they serve a more fundamental purpose as reminders of the incredible adaptability of life. What could be more alien to us than organisms surviving in an anoxic environment thousands of meters below the surface of the ocean, living off energy obtained through geothermal and not solar processes.[12]

And as Smith continues:

The discovery of chemosynthetic communities in the deep sea, starting with hydrothermal vents in 1977, is possibly one of the most significant biological discoveries of the late 20th century. It had the effect of restructuring the general view on the importance of the sun and has been an area of significant research activity for the last 30 y[ea]r[s] ... While often characterized as communities in isolation from the sun, much of the life on hydrothermal vents, cold seeps, whale and wood falls could not and would not exist in its absence. The underlying point however, that there exists life which does not require the sun at these sites, remains true. A small portion of organisms in the deep sea make use of anaerobic chemosynthesis which does not have the same free oxygen requirements as the aerobic version. Even as the majority of life on earth would vanish without the sun, there exists the continued potential for a small group of lonely microorganisms to persist deep down in the perpetual darkness of our deepest seas.[13]

In effect, the possibility of a complex biosphere thriving in the dark is no longer science fiction. And of course, it means that biospheres beyond the Earth may also thrive in the dark. The post-1977 discoveries mean that the conventional view defended by Wald, the mantra that everyone learned at school in the 1950s and 1960s that all carbon-based life on Earth depends on the Sun, is certainly invalidated.

But while the conventional pre-*Alvin* paradigm—that all life on Earth depends on photosynthesis—has been invalidated after *Alvin,* it is still true that all complex organisms on Earth, including the metazoan vent biota—crabs, tube worms, and fish (which live by oxidizing the reduced carbon organics manufactured by the chemosynthetic organisms)—as well as bees and birds—depend on oxidations using *molecular oxygen as the terminal oxidant* to generate their metabolic energy (ATP), and hence depend on photosynthesis and in turn on the light of the Sun. The *mantra* that *all complex life* necessarily must use oxidations for metabolic energy (ATP) still holds and indeed has been reinforced after *Alvin.* Wald may have been wrong in claiming that life cannot originate or survive in the dark, but his claim that organisms that do not use oxidations "never amounted to much"[14] is amply confirmed by the biota round the vents.

In summing up his overview on microbial metabolic diversity in *Science,* Norman Pace commented:

> Textbooks generally portray only a small part of the global distribution of life, the part that is immediately dependent on either the harvesting of sunlight or the metabolism of the decay products of photosynthesis... [however] lithotrophic metabolism... is more widespread phylogenetically and geographically than is either phototrophy or organotrophy. The lithotrophic biosphere potentially extends kilometres into the crust of the Earth, an essentially unknown realm. These considerations may indicate that lithotrophy contributes far more to the biomass of Earth than currently thought.
>
> Part of that lithotrophic biomass is in microhabitats all around us, usually away from light and oxygen. It is not necessary to look far to find such environments: the rumens of cattle and the guts of termites and humans, for example, are significant sources of methane,

a signature of hydrogen metabolism. Most life that depends on inorganic energy metabolism, however, probably is in little-known environments, based on poorly understood geo-chemistries...

... Nonetheless it seems possible that much, perhaps most, *of the* biomass *on Earth is subterranean, a biological world based on lithotrophy.*[15]

Much subsequent research has provided further support for the notion that the subterranean denizens of the dark may make up a very substantial proportion of Earth's biomass.[16] Recent research reviewed in *New Scientist* by Catherine Brahic confirms Pace's inference revealing that the rocks beneath the oceans could be home to the largest population of prokaryotes on Earth[17] and that the combined undersea biomass could be equivalent to that of all the plants on Earth.[18] Clearly, whatever might have been the mantra in the mid-twentieth century, "light eating" is just one life strategy among a vast diversity of alternatives and "light eaters" may comprise only a relatively small fraction of the total biomass of life on Earth.

But although our light-eating lifestyle and dependence on the oxidation of what Pace terms "the decay products of photosynthesis" may be only *one lifestyle out of a vast suite of alternative strategies available to carbon-based life,* dramatically manifest in the warm surrounds of the hydrothermal vents, it is exceptional and unique in one very significant way. It is the *only lifestyle on Earth,* and as far as we can tell the only lifestyle anywhere in the universe, *able to supply complex, energy-hungry, carbon-based organisms like ourselves with sufficient energy to satisfy our metabolic needs.* Only light-eating can sustain intelligent organisms capable of developing a high civilization. Without the Sun there would be no bumblebees drinking nectar from a flower, no fire, no metallurgy, no chemistry, no technology, and no humans planning a trip to Mars.

The discovery and documentation, post-*Alvin,* of an astonishing menagerie of metabolically diverse microbes and a universe of extremophile species,[19] able to thrive in hostile anoxic environments very different from those with which we are familiar, has greatly extended the boundaries of carbon-based life. It is now clear that a world of carbon-based life peopled by a vast assemblage of organisms can thrive without the fine-tuning of starlight for photosynthesis, without the fine tuning of the atmosphere to let through the 'right' light, without the fine tuning for high-acuity vision, without the fine tuning for terrestrial plants. We now know there can be a cosmos replete with carbon-based life; yet, without the additional elements of fine tuning for us energy-hungry aerobes, it would be devoid of complex, advanced, carbon-based organisms remotely comparable with ourselves. The recent discovery of subsurface organics in Martian sediments that are three billion years old[20] suggests that there may have been carbon-based Martian life in the remote past. But even if exotic carbon-based life still lurks in the sands of Mars, lacking the right atmosphere, lacking oxygen and the protective ozone shield, it can only be the most primitive sort of unicellular life, imprisoned forever in the dark subsurface sands.

The significance of the extraordinary ensemble of fitness in nature for our unique, light-dependent lifestyle has been enormously enhanced by the extension of the boundaries. For, in the context of the multiverse of alternative lifestyles, the fact that nature exhibits so many elements of fine tuning for our lifestyle—for advanced, aerobic, terrestrial, carbon-based organisms possessed with high intelligence and sight—is surely of enormous significance, supportive of a profound anthropocentric bias in the nature of things. The fact that the properties of water tell the same story (see *The Wonder of Water*), as does the fitness of nature for fire-making by organisms of our biological design (see *Fire-Maker*), greatly confirms this anthropocentric bias, and further challenges the post-Copernican notion that there is no special place for beings of our biology in the world order.

Appendix B. Fermentation and Cellular Respiration

> It is difficult to overestimate the degree to which the invention of cellular respiration [oxidation] released the forces of living organisms... No organism that relies wholly on fermentation has ever amounted to much... Respiration [oxidation] used the material of organisms [reduced carbon fuels] with such enormously greater efficiency as for the first time to leave something over... To use an economic analogy... respiration provided them with capital. It is mainly this capital that they invested in the great enterprise of organic evolution.
>
> —George Wald (1954), *Scientific American*[1]

Fermentation

Virtually all organisms possess an anaerobic metabolic pathway called *glycolysis or fermentation* (referred to in the Wald passage quoted above) which generates metabolic energy in the form of ATP from reactions which can occur in the absence of free oxygen.[2] Fermentation (or glycolysis) reactions convert one molecule of sugar into two molecules of pyruvic acid. Pyruvic acid is readily converted to ethanol [alcohol]—hence the term fermentation.

$$\underset{\text{sugar}}{C_6H_{12}O_2} \Rightarrow \underset{\text{pyruvic acid}}{CH_3COCOOH} + 2\ ATP$$

Although the reactions involved in glycolysis do not involve free oxygen (O_2), many are classed by chemists as oxidations because they involve molecular rearrangements which change the redox state[3] of one or more of the reactants towards a more oxidized state. (Note that the proportion of oxygen to carbon and hydrogen is greater in pyruvate than in sugar.) So because of the near universality of glycolysis, in one sense all life on earth utilizes oxidation, or at least a form of oxidation involving molecular rearrangements. For every sugar molecule converted to pyruvic acid, two energy-rich ATP molecules are manufactured. A complete description of glycolysis is given in most texts of biochemistry.

Cellular Respiration

THIS TERM generally refers to the aerobic process whereby partially oxidized products of glycolysis, such as pyruvate and other products of metabolism, are fully oxidized to CO_2 and H_2O, releasing energy which is used to synthesize ATP molecules in the mitochondria.

$$\text{Pyruvate and other partially oxidized metabolites} + O_2$$
$$\Downarrow$$
$$CO_2 + H_2O + \sim 30 \text{ ATP}$$

The complete oxidation of sugar to CO_2 and H_2O (glycolysis plus cellular respiration) yields approximately ten times the number of ATP molecules produced by glycolysis alone (hence Wald's remarks). The mechanisms involved in the generation of ATP in respiration are complex, involving electron transport chains (ETC) situated in the inner mitochondrial membrane, through which the electrons released during metabolism flow down an electrochemical gradient terminating in the reduction of oxygen (the terminal oxidant in aerobic respiration) to H_2O. (The fitness of the transitional metals to form the electrical conducting wires in ETCs will be discussed in a subsequent monograph.) The energy dissipated as the electrons flow down the ETC is used to pump protons across the inner mitochondrial membrane into the intermembrane space, from which they flow back through the membrane, driving the synthesis of ATP by the enzyme ATP synthase which straddles the inner mitochondrial membrane. (For a detailed description of cellular respiration see any major text of biochemistry.[4])

Strictly speaking, the term cellular respiration, although in common usage referring to the aerobic type which occurs in the mitochondria of all advanced aerobic organisms, is not the only type of cellular respiration. Many unicellular organisms (such as many of those which thrive around the hydrothermal vents; see Appendix A) utilize electron acceptors other than oxygen as the terminal oxidant. Methanogens use CO_2 which is reduced to methane (CH_4). Other microbes reduce sulphate (SO_4) to sulphide.[5] Some even "breathe metals,"[6] reducing ferric ions to ferrous ions.[7] What all these alternative "oxidants" have in common is that their reduction generates far less energy than the reduction of oxygen to water. All anaerobes using alternative oxidants to free oxygen are simple unicellular life forms.

Glycolysis or fermentation is therefore a very different means of ATP production from cellular respiration. Further, the two mechanisms are carried out in different compartments in the cell. The reactions of glycolysis occur in the cytoplasm of eukaryotic cells, whereas the reactions involved in cellular respiration occur in the mitochondria.

As metabolically active aerobes, we are critically dependent on the generous supply of ATP provided by cellular respiration using oxygen as the terminal oxidant. However, many organisms can get by with far less ATP. Surprisingly, some vertebrates, including some species of carp (e.g., the Crucian carp) and North American fresh-water turtles, can survive at low temperatures for several months during winter under the ice, generating all the ATP they need from glycolysis alone.[8]

Such vertebrates possess several specialized biochemical adaptations which assist them in surviving during periods when they are dependent on glycolysis for their energy needs. Although the amount of energy produced during glycolysis is only a tenth of that produced in aerobic conditions, when the carbon and hydrogen atoms in sugar or fat are fully oxidized to water and carbon dioxide, these cold-blooded organisms have massively reduced energy requirements at near-zero Celsius during the winter months, even compared with a warm-blooded human at rest.[9] Nonetheless, even these uniquely adapted vertebrate species, despite their special biochemical adaptations, can only survive without oxygen for a few days at room temperatures.[10]

With metabolic rates much greater than those of cold-blooded vertebrates, such an option is not available for warm-blooded mammals like ourselves. We *need* 250 milliliters of oxygen *every minute* without ceasing throughout our lives, and up to five liters per minute during strenuous activity.[11]

Although it has recently been reported that a primitive metazoan (multicellular species) dredged up from the sediments at the bottom of the Mediterranean can survive and complete its life cycle in an anoxic environment,[12] all other known metazoan (multicellular) organisms, including the Crucian carp and fresh-water turtles, depend ultimately on the utilization of molecular oxygen and can only complete their reproductive cycles in aerobic conditions. As well, the great majority of complex, one-celled, nucleated organisms (eukaryotes) are also oxygen-users, with only a few exceptions. And of course, all green plants and algae are aerobic, containing mitochondria and deriving metabolic energy (ATP) from cellular respiration. (They also derive energy, or ATP, for organic synthesis directly from sunlight. See the discussion in Chapter 5.) The handful of unicellular eukaryotes that are anaerobic are either parasites or obligate symbionts of multicellular organisms.[13]

Endnotes

1. The Miracle of Sunlight

1. Carl Sagan, *Cosmos* (New York: Ballantine Books, 1980), 198–199. Emphasis added.

2. Richard Cohen, *Chasing the Sun: The Epic Story of the Star That Gives Us Life* (New York: Simon and Schuster, 2010), Chapter 3.

3. Ibid. The Sphinx has faced the rising Sun on the morning of the spring or vernal equinox on the Giza plateau every year since it was built during the reign of Pharaoh Khafre about 2500 B. C. The March equinox marks the moment the Sun crosses the celestial equator—the imaginary line in the sky above the Earth's equator—from south to north (and *vice versa* in September) marking the return of the sun and the first day of spring.

4. Ibid.

5. Cohen, Chapter 3. There is a stone circle twenty-five meters across on the top of Medicine Mountain in the Big Horn range in Wyoming, in which one of the spokes is aligned to the summer solstice sunrise within one-third of a degree. See J. A. Eddy, "Astronomical Alignment of the Big Horn Medicine Wheel," *Science* 184, no. 4141 (June 7, 1974): 1035–1043. https://doi.org/10.1126/science.184.4141.1035.

6. *Ivan Šprajc, "Astronomical Alignments at Teotihuacan, Mexico," Latin American Antiquity 11, no. 4 (2000): 403–415; for more information, also see* "Teotihuacan," *Wikipedia: The Free Encyclopedia,* Wikimedia Foundation, January 24, 2018, accessed January 31, 2018, https://en.wikipedia.org/wiki/Teotihuacan.

7. Sprajc (2000); Cohen (2010) on page 48 describes how at the setting and rising of the Sun on the spring and autumn equinox, shadows cast by the corners of the pyramid known as El Castillo, in the Mayan city of Chichen Itza, casts the outline of a feathered serpent apparently crawling down the pyramid. See A. F. Aveni, S. L. Gibbs, and H. Hartung, "The Caracol Tower at Chichen Itza: An Ancient Astronomical Observatory?" *Science* 188, no. 4192 (June 6, 1975): 977–985. https://doi.org/10.1126/science.188.4192.977.

8. *Encyclopedia Britannica,* s.v. "Sun worship," accessed January 31, 2018, https://www.britannica.com/topic/sun-worship.

9. Ibid.; Cohen, Chapter 1.

10. Cohen, Chapter 1.

11. *Encyclopedia Britannica*, s.v. "Sun worship," https://www.britannica.com/topic/sun-worship; "Helios," *Wikipedia: The Free Encyclopedia*, The Wikimedia Foundation, January 30, 2018, accessed January 31, 2018, https://en.wikipedia.org/wiki/Helios.

12. Bob Berman, *The Sun's Heartbeat: And Other Stories from the Life of the Star That Powers Our Planet* (New York: Back Bay Books, 2012), 7.

13. William J. Broad, *The Universe Below: Discovering the Secrets of the Deep Sea* (New York: Simon & Schuster, 1997), 109: "The ability of some microbes to live off chemicals rather than light was well known before the discovery of the deep vents, but it was a major revelation to learn that highly complex ecosystems were powered by this principle [chemosynthesis]—that we and all the other light-eaters of Earth shared our planet with an alien horde that thrived in total darkness."

14. P. R. Renne, A. L. Deino, F. J. Hilgen, K. F. Kuiper, D. F. Mark, W. S. Mitchell, L. E. Morgan, R. Mundil, and J. Smit, "Time Scales of Critical Events Around the Cretaceous-Paleogene Boundary," *Science* 339, no. 6120 (2013): 684–687. Available online at: https://eps.harvard.edu/files/eps/files/renne.kt_.science.2013.pdf.

15. Broad, *The Universe Below*, 109.

2. The Light of Life

1. George Wald, "Light and Life," *Scientific American* 201, no. 4 (1959): 92–108.

2. William J. Broad, *The Universe Below: Discovering the Secrets of the Deep Sea* (New York: Simon & Schuster, 1997), 109.

3. Geoffrey K. Vallis, *Climate and the Oceans*, Princeton Primers in Climate (Princeton: Princeton University Press, 2012), Glossary, page 218. Vallis writes: "Without this effect [the greenhouse], Earth's surface would have a temperature of about 255 K (-18°C), about 33 degrees lower than it actually is [i.e., 15°C]."

4. John F. B. Mitchell, "The 'Greenhouse' Effect and Climate Change," *Reviews of Geophysics* 27, no. 1 (1989): 115–139; see Figure 2.1. Regarding black bodies, Mitchell notes: "The distribution of energy emitted with wavelength is a function of the temperature of the emitted: the hotter the emitter, the shorter the wavelength of peak emission. Thus the sun, which has a surface temperature of 6000K, emits most radiation in the range of 0.2 to 4 microns (including ultraviolet, visible, and near infra red), whereas the Earth at 255K emits mainly in the range of 4–100 microns." See *Encyclopaedia Britannica*, s.v. "Continuous spectra of electromagnetic radiation," accessed February 13, 2018, https://www.britannica.com/science/electromagnetic-radiation/Continuous-spectra-of-electromagnetic-radiation.

5. Mitchell (1989).

6. Ibid.

7. "Ultraviolet Astronomy," *Wikipedia: The Free Encyclopedia*, The Wikimedia Foundation, December 28, 2017, accessed January 11, 2018, https://en.wikipedia.org/wiki/Ultraviolet_astronomy. From the article: "The ultraviolet Universe looks quite different from the familiar stars and galaxies seen in visible light. *Most stars are actually relatively cool objects emitting much of their electromagnetic radiation in the visible or near-infrared part of the spectrum.* Ultraviolet radiation is the signature of hotter objects, typically in the early and late stages of their evolution." [my emphasis]

8. "Stellar Classification," *Wikipedia: The Free Encyclopedia,* The Wikimedia Foundation, December 30, 2017, accessed January 11, 2018, https://en.wikipedia.org/wiki/Stellar_classification.

9. Carl Sagan, *Cosmos* (New York: Random House, 1980), 243 (Chapter IX); "Stellar Classification," *Wikipedia,* https://en.wikipedia.org/wiki/Stellar_classification.

10. Carlos A. Bertulani, *Nuclei in the Cosmos* (Hackensack, NJ: World Scientific, 2013), Chapter 1, table on page 10.

11. Ibid., Chapter 1.

12. Ibid.

13. Ibid.

14. "Stellar Classification," *Wikipedia,* https://en.wikipedia.org/wiki/Stellar_classification.

15. "Colors, Temperatures, and Spectral Types of Stars," *John A. Dutton e-Education Institute,* PennState, 2017, accessed January 11, 2018, https://www.e-education.psu.edu/astro801/content/l4_p2.html.

16. Hellmut Fritzsche and Melba Phillips, *Encyclopaedia Britannica,* 15th edition, s.v. "Electromagnetic radiation" (Chicago: Encyclopaedia Britannica, 1994), vol. 18, page 197. Also available online at Hellmut Fritzsche and Melba Phillips, *Encyclopaedia Britannica,* s.v. "Electromagnetic radiation: General considerations," accessed February 13, 2018, https://www.britannica.com/science/electromagnetic-radiation#toc59189.

17. Ibid.; C. R. Nave, "The Interaction of Radiation with Matter," *HyperPhysics,* Module 3, 2016, accessed January 11, 2018, http://hyperphysics.phy-astr.gsu.edu/hbase/mod3.html.

18. Michael Denton, *Nature's Destiny: How the Laws of Biology Reveal Purpose in the Universe* (New York: Free Press, 1998), 51.

19. One light year is approximately 10^{12} Km; the Andromeda galaxy is 2.5 million light years away ["The Galaxy Next Door," *Galex,* NASA, May 15, 2012 (updated August 7, 2017), accessed January 11, 2018, https://www.nasa.gov/mission_pages/galex/pia15416.html] and assuming a playing card to be 0.5 mm thick this gives a distance of 5 million light years.

20. *Encyclopaedia Britannica,* s.v. "Electromagnetic Radiation," and online at *Encyclopaedia Britannica,* s.v. "Electromagnetic radiation: General considerations," https://www.britannica.com/science/electromagnetic-radiation#toc59189.

21. Ibid.

22. George Wald, "Life and Light," *Scientific American* 201, no. 4 (1959): 92–108; Campbell, *Energy and the Atmosphere,* 1–2; Nave, "The Interaction of Radiation with Matter," http://hyperphysics.phy-astr.gsu.edu/hbase/mod3.html.

23. Wald, "Life and Light."

24. Nave, "The Interaction of Radiation with Matter," http://hyperphysics.phy-astr.gsu.edu/hbase/mod3.html.

25. Wald, "Life and Light."

26. Ibid.

27. Ibid.

28. "Triton (moon)," *Wikipedia: The Free Encyclopedia*, The Wikimedia Foundation, January 8, 2018, accessed January 11, 2018, https://en.wikipedia.org/wiki/Triton_(moon).

29. H. Eyring, R. P. Boyce, and J. D. Spikes, "Thermodynamics of Living Systems," in *Comparative Biochemistry*, vol. 1, ed. M. Florkin and H. S. Mason (New York: Academic Press, 1960), 60–62.

30. Ibid.

31. R. E. D. Clarke, *The Universe: Plan or Accident?* (London: Paternoster Press, 1961), 59.

32. L. J. Rothschild and R. L. Mancinelli, "Life in Extreme Environments," *Nature* 409, no. 6823 (February 22, 2001): 1092–1101. https://doi.org/10.1038/35059215.

33. Ken Takai, Kentaro Nakamura, Tomohiro Toki, Urumu Tsunogai, Masayuki Miyazaki, Junichi Miyazaki, Hisako Hirayama, Satoshi Nakagawa, Takuro Nunoura, and Koki Horikoshi, "Cell Proliferation at 122 Degrees C and Isotopically Heavy CH4 Production by a Hyperthermophilic Methanogen under High-Pressure Cultivation," *Proceedings of the National Academy of Sciences of the United States of America* 105, no. 31 (August 5, 2008): 10949–54. doi:10.1073/pnas.0712334105.

34. M. J. Denton, *The Wonder of Water* (Seattle: Discovery Institute Press, 2017), 163.

35. Luke Mastin, "Timeline of the Big Bang," *The Physics of the Universe*, 2009, accessed January 11, 2018, http://www.physicsoftheuniverse.com/topics_bigbang_timeline.html; the Planck temperature is 10^{32}K.

36. E. M. Burbidge, G. R. Burbidge, W. A. Fowler, and Fred Hoyle, "Synthesis of the Elements in Stars," *Reviews of Modern Physics* 29 (1957): 547–650.

37. S. Mitta, *Cambridge Encyclopaedia of Astronomy* (London: Jonathan Cape, 1977), 128; Dedra Forbes, "Temperature at the Center of the Sun," *The Physics Factbook*, 1997, accessed January 11, 2018, http://hypertextbook.com/facts/1997/DedraForbes.shtml.

38. Denton, *The Wonder of Water*, 163.

39. Mitchell (1989).

40. The authors of a recent paper [Dimitris J. Panagopoulos, Olle Johansson, and George L. Carlo, "Evaluation of Specific Absorption Rate as a Dosimetric Quantity for Electromagnetic Fields Bioeffects," ed. Nils Cordes, *PLoS ONE* 8, no. 6 (June 4, 2013): e62663, https://doi.org/10.1371/journal.pone.0062663] point out: "[T]here is already a large and constantly increasing number of studies indicating that environmental man-made EMFs can produce severe biological alterations such as DNA damage without heating the biological tissue… This can take place through non-thermal mechanisms that involve direct changes in intracellular ionic concentrations or changes in enzymatic activity… DNA damage may lead to cancer, neurodegenerative diseases, reproductive declines, or even heritable mutations. Brain tumors, decrease in reproductive capacity, or symptoms reported as "microwave syndrome" (headaches, memory loss, fatigue, etc.), are observed among people exposed to mobile telephony radiation during recent years… Recently the International Agency for Research on Cancer (IARC) has classified RF/microwave EMFs as "possibly carcinogenic to humans." Yet the level of radiation emitted by these devices is very small and insufficient to induce temperature increases: "[I]n real

exposure conditions as e.g. in the case of a GSM mobile phone during normal "talk" operation the average power density even in contact with the antenna hardly exceeds 0.2–0.3 mW/cm^2 and *does not induce temperature increases at a 0.05 C level.*" In other words, *the deleterious effects are not caused by internal heating of body tissues but because of the impact of the radiation directly on bio-matter.*

Other recent publications report similar concerns: C. Rougier, A. Prorot, P. Chazal, P. Leveque, and P. Leprat, "Thermal and Nonthermal Effects of Discontinuous Microwave Exposure (2.45 Gigahertz) on the Cell Membrane of Escherichia Coli," *Applied and Environmental Microbiology* 80, no. 16 (August 15, 2014): 4832–41, doi:10.1128/AEM.00789-14; O. V. Furtado-Filho, "Ultra-High-Frequency Electromagnetic Radiation and Reactive Species in Mammals," in *Microwave Effects on DNA and Proteins*, ed. Chris D. Geddes (Cham, Switzerland: Springer International Publishing, 2017), 249–274; R. N. Kostoff and C. G. Y. Lau, "Modified Health Effects of Non-ionizing Electromagnetic Radiation Combined with Other Agents Reported in the Biomedical Literature," in the aforementioned *Microwave Effects on DNA and Proteins*, 97–158; Jitendra Behari and Tanu Jindal, "Microwave Effects on DNA," again in *Microwave Effects on DNA and Proteins*, 67–98.

In short : To heat the earth substituting microwave radiation for IR would require radiation fluxes at least sufficient to raise the temperature of water and also body tissues (which are 70% water) i.e., greater than those generated artificially in devices such as mobiles, emitting about 0.3 mW/cm2, which do not cause heating but do appear to have deleterious effects. To achieve the same energy flux as that of current IR levels and to heat the earth to an equivalent degree it would be necessary to impose on earth a continuous microwave radiation flux sufficient to heat water by several degrees C.

41. Mitchell (1989); K. L. Coulson, *Solar and Terrestrial Radiation* (New York: Academic Press, 1975), see Chapter 3, page 40, Figure 3.1.

42. Fred Hoyle, *Home Is Where the Wind Blows: Chapters from a Cosmologist's Life* (Mill Valley, CA: University Science Books, 1994), Chapter 18.

43. Wald, "Light and Life." Emphasis added.

44. T. H. Goldsmith, "Photoreception and Vision," in *Comparative Animal Physiology*, ed. C. L. Prosser (Philadelphia: Saunders, 1973), 577–632, in particular page 577.

45. Campbell, *Energy and the Atmosphere*, 1.

46. Ibid., 1–2.

47. Ibid. Emphasis added.

48. Johnjoe McFadden and Jim Al-Khalili, *Life on the Edge: The Coming of Age of Quantum Biology* (New York: Broadway Books [Crown Publishing], 2014), Chapter 4. Also see McFadden, "It Seems Life Really Does Have a Vital Spark: Quantum Mechanics," http://www.abc.net.au/news/2016-02-08/mcfadden-it-seems-life-really-does-have-a-vital-spark/7148448; Frank Trixler, "Quantum Tunnelling to the Origin and Evolution of Life," *Current Organic Chemistry* 17, no. 16 (July 1, 2013): 1758–1770. https://doi.org/10.2174/13852728113179990083. And see video discussion at World Science Festival, "Quantum Biology: The Hidden Nature of Nature," September 17, 2015, YouTube, accessed May 9, 2018, https://www.youtube.com/watch?v=ADiql3FG5is&t=297s.

49. Richard P. Feynman, Robert B. Leighton, and Matthew L. Sands, *The Feynman Lectures on Physics*, New Millennium ed., vol. 3 (New York: Basic Books, 2011, first published in 1966), see opening paragraph. E-copy available at http://www.feynmanlectures.caltech.edu/III_01.html.
50. Richard Phillips Feynman, *QED: The Strange Theory of Light and Matter*, Princeton Science Library (Princeton, N.J: Princeton University Press, 1988), 10.
51. See, for example, this ninety-minute video recorded at the late 2015 World Science Festival, of a conversation between Paul Davies, Seth Lloyd (who works on quantum computing at MIT), and Thorsten Ritz (who works on quantum magnetic detection in avian eyes at the University of California, Irving): "Quantum Biology: The Hidden Nature of Nature": https://www.youtube.com/watch?v=ADiql3FG5is&t=297s. Many other similar discussions can be found by searching the web.
52. McFadden and Al-Khalili, *Life on the Edge*, Chapter 1.
53. Trixler (2013).
54. It is possible to envisage how this happens with a simple and somewhat crude model, but one which captures something of the phenomenon: Take a tray and fill it with about three centimeters of water. Then place several stones in an approximate line across the middle of the tray so that there is no direct straight line path through the center of the tray from one end to the other. Obviously no particle propelled from one end of the tray can ever reach the other end of the tray because its path is blocked by the barrier of stones But create a wave at one end of the tray and it will pass through the channels between the stones no matter how convoluted and reach the other end of the tray.
55. Trixler (2013). See also "Quantum Tunnelling," *Wikipedia: The Free Encyclopedia*, The Wikimedia Foundation, January 9, 2018, accessed January 11, 2018, https://en.wikipedia.org/wiki/Quantum_tunnelling.
56. Trixler (2013), 1761. "Despite the hot environment of stellar interiors, the nucleosynthesis process requires quantum tunneling in order to occur because of insufficient thermal energy for overbarrier fusion reactions."
57. Trixler (2013).
58. Trixler (2012).
59. McFadden and Al-Khalili, *Life on the Edge*, 91–92.
60. Trixler (2012); McFadden and Al-Khalili, *Life on the Edge*, Chapter 3.
61. Edward J. O'Reilly and Alexandra Olaya-Castro, "Non-Classicality of the Molecular Vibrations Assisting Exciton Energy Transfer at Room Temperature," *Nature Communications* 5 (January 9, 2014). https://doi.org/10.1038/ncomms4012. As the authors comment: "Our results therefore suggest that investigation of the non-classical properties of vibrational motions assisting excitation and charge transport, photoreception and chemical sensing processes could be a touchstone for revealing a role for non-trivial quantum phenomena in biology."
62. McFadden and Al-Khalili, 123.
63. Trixler (2013). Emphasis added.
64. Gordon B. Haxel, James B. Hedrick, and Greta J. Orris, "Rare Earth Elements—Critical Resources for High Technology," U. S. Geological Survey, Fact Sheet

087-02, *USGS*, November 20, 2002, accessed June 6, 2018, http://pubs.usgs.gov/fs/2002/fs087-02/.

65. Hellmut Fritzsche and Melba Phillips, *Encyclopaedia Britannica*, 15th edition, s.v. "Electromagnetic radiation," (Chicago: Encyclopaedia Britannica, 1994), vol. 18, see page 197. Also available online at Hellmut Fritzsche and Melba Phillips, *Encyclopaedia Britannica*, s.v. "Electromagnetic radiation: General considerations," accessed February 13, 2018, https://www.britannica.com/science/electromagnetic-radiation#toc59189.
66. Philip Charles Nelson, Sarina Bromberg, Ann Hermundstad, and Jesse M. Kinder, *From Photon to Neuron: Light, Imaging, Vision* (Princeton, NJ: Princeton University Press, 2017), Chapter 8.
67. A. Sayre, *Rosalind Franklin and DNA* (New York: Norton, 1975).
68. "Ultraviolet," *Wikipedia: The Free Encyclopedia*, Wikimedia Foundation, January 4, 2018, accessed January 12, 2018, https://en.wikipedia.org/wiki/Ultraviolet.
69. Carl Sagan's first published use of the expression was apparently in *The Cosmic Connection: An Extraterrestrial Perspective* (New York: Anchor Press/Doubleday, 1973), pp. 189–190.
70. William J. Broad, *The Universe Below: Discovering the Secrets of the Deep Sea* (New York: Simon & Schuster, 1997), 109.
71. Campbell (1977), 1–2.

3. Letting the Light In

1. Carl Sagan, *Cosmos* (New York: Ballantine Books, 1980), 80.
2. See "Electromagnetic Radiation," *Encyclopaedia Britannica*, 15th edition, vol. 18 (Chicago: Encyclopaedia Britannica, 1994), p. 200; see Figure 5.
3. "Electromagnetic Radiation," in op. cit., vol. 18, p. 198; see Figure 3; copy available at https://www.britannica.com/science/electromagnetic-radiation/Microwaves#toc59184.
4. IR radiation is absorbed within ten centimeters of surface. See "Light in the Ocean," *Exploring Our Fluid Earth*, University of Hawai'i, 2008, accessed January 12, 2018, https://manoa.hawaii.edu/exploringourfluidearth/physical/ocean-depths/light-ocean, and John L. Daly, "The Top of the World: Is the North Pole Turning to Water?" *Still Waiting for Greenhouse*, August 15, 2008, https://www.john-daly.com/polar/arctic.htm.
5. For absorption by liquid water, see "Electromagnetic Radiation," cited above, p. 198, Figure 3. See also Figure 5.4 in "Electromagnetic Absorption by Water," *Wikipedia: The Free Encyclopedia*, Wikimedia Foundation, October 10, 2017, https://en.wikipedia.org/wiki/Electromagnetic_absorption_by_water.
6. For absorption of water vapor (and other atmospheric gases) see "Absorption band," *Wikipedia: The Free Encyclopedia*, Wikimedia Foundation, October 14, 2017, accessed January 12, 2018, https://en.wikipedia.org/wiki/Absorption_band; see also the above-cited "Electromagnetic Radiation."
7. Stephen G. Warren, Richard E. Brandt, and Thomas C. Grenfell, "Visible and near-Ultraviolet Absorption Spectrum of Ice from Transmission of Solar

Radiation into Snow," *Applied Optics* 45, no. 21 (July 20, 2006): 5320. doi:10.1364/AO.45.005320. The authors comment: "The general features of the spectrum are well known... Ice exhibits strong absorption in the ultraviolet (UV) at wavelength 170 nm. With increasing wavelength, the absorption becomes extremely weak in the visible, with a minimum near 400 nm." See also Stephen G. Warren and Richard E. Brandt, "Optical Constants of Ice from the Ultraviolet to the Microwave: A Revised Compilation," *Journal of Geophysical Research* 113, no. D14 (July 31, 2008). doi:10.1029/2007JD009744.

8. Fritzsche and Phillips, "Microwaves: Visible Radiation," https://www.britannica.com/science/electromagnetic-radiation/Microwaves#toc59184.

9. John F. B. Mitchell, "The 'Greenhouse' Effect and Climate Change," *Reviews of Geophysics* 27, no. 1 (1989): 115–139. This paper includes excellent information on the atmospheric absorption of IR and diagrams showing the bell-shaped spectrums of EM radiation emitted by the Sun and Earth. While the Sun's EM emission peaks at about 0.5 microns close to the middle of the visual band, the Earth's peaks at about fifteen microns.

10. From David R. Williams, "Moon Fact Sheet," *Lunar and Planetary Science*, NASA, July 3, 2017, accessed January 12, 2018, https://nssdc.gsfc.nasa.gov/planetary/factsheet/moonfact.html. The diurnal temperature range on the Moon (at the equator) is from 95 K [-178°C] to 390 K [116°C]. See also Hellmut Fritzsche and Melba Phillips, "Continuous spectra of electromagnetic radiation: Greenhouse effect on the atmosphere," *Encyclopaedia Britannica*, accessed February 13, 2018, https://www.britannica.com/science/electromagnetic-radiation/Continuous-spectra-of-electromagnetic-radiation. See also Fraser Cain, "Temperature of the Moon," *Universe Today*, October 13, 2008, https://www.universetoday.com/19623/temperature-of-the-moon/, where the author writes: "Are you planning a trip to the Moon and you're wondering what kinds of temperature you might experience. Well, you're going to want to pack something to keep you warm, since the temperature of the Moon can dip down to -153°C during the night. Oh, but you're going to want to keep some cool weather clothes too, since the temperature of the Moon in the day can rise to 107°C. Why does the Moon's temperature vary so widely? It happens because the Moon doesn't have an atmosphere like the Earth. Here on Earth, the atmosphere acts like a blanket, trapping heat. Sunlight passes through the atmosphere, and warms up the ground. The energy is emitted by the ground as infrared radiation, but it can't escape through the atmosphere again easily so the planet warms up. Nights are colder than days, but it's nothing like the Moon. There's another problem. The Moon takes 27 days to rotate once on its axis. So any place on the surface of the Moon experiences about 13 days of sunlight, followed by 13 days of darkness. So if you were standing on the surface of the Moon in sunlight, the temperature would be hot enough to boil water. And then the Sun would go down, and the temperature would drop 250 degrees in just a matter of moments."

11. Mitchell (1989).

12. Geoffrey K. Vallis, *Climate and the Oceans*, Princeton Primers in Climate (Princeton: Princeton University Press, 2012), see Glossary, page 218. Vallis writes: "Without this effect [the greenhouse], Earth's surface would have a temperature of about 255K (-18°C), about 33 degrees lower than it actually is." [The global mean temperature is 15°C.]

13. William Cotton, *Human Impacts on Weather and Climate* (Cambridge: Cambridge University Press, 2006). Cotton comments on page 156: "Little absorption is evident in the region called the atmospheric window between 8 and 14 μm." See also Mitchell (1989), Figure 2; also see NASA image of atmospheric absorption shown at "Greenhouse Effect," *Wikipedia: The Free Encyclopedia*, The Wikimedia Foundation, December 16, 2017, accessed January 12, 2018, https://en.wikipedia.org/wiki/Greenhouse_effect.
14. Mitchell.
15. Ibid.
16. J. N. Maina, "Structure, Function and Evolution of the Gas Exchangers: Comparative Perspectives," *Journal of Anatomy* 201, no. 4 (October 2002): 281–304.
17. David C. Catling, Christopher R. Glein, Kevin J. Zahnle, and Christopher P. McKay, "Why O_2 Is Required by Complex Life on Habitable Planets and the Concept of Planetary 'Oxygenation Time,'" *Astrobiology* 5, no. 3 (June 2005): 415–38. https://doi.org/10.1089/ast.2005.5.415.
18. "Ozone," *Wikipedia: The Free Encyclopedia*, The Wikimedia Foundation, May 6, 2018, accessed June 6, 2018, https://en.wikipedia.org/wiki/Ozone.
19. Lawrence Henderson, *The Fitness of the Environment* (New York: The MacMillan Company, 1913), see Chapter 4, pages 138–139. In Henderson's words, "Because of its solubility [and]... because of the reservoir of the atmosphere... Its occurrence is universal and its mobility a maximum." E-copy available at https://archive.org/stream/cu31924003093659#page/n157/mode/2up.
20. A. E. Needham, *Uniqueness of Biological Materials* (Oxford; London; Edinburgh: Pergamon Press, 1965), 35.
21. Douglas Drysdale, *An Introduction to Fire Dynamics*, 3rd ed. (Chichester, West Sussex: Wiley, 2011), 377; Clayton Huggett, "Habitable atmospheres which do not support combustion," *Flame and Combustion* 20, no. 1 (February 1973): 140–142.
22. Drysdale; Huggett, 140–142.
23. "Carbon Dioxide through Geologic Time," *Climage Change: Past and Future*, University of California, San Diego, 2002, accessed January 12, 2018, http://earthguide.ucsd.edu/virtualmuseum/climatechange2/07_1.shtml.
24. Joseph G. Allen, Piers MacNaughton, Usha Satish, Suresh Santanam, Jose Vallarino, and John D. Spengler, "Associations of Cognitive Function Scores with Carbon Dioxide, Ventilation, and Volatile Organic Compound Exposures in Office Workers: A Controlled Exposure Study of Green and Conventional Office Environments," *Environmental Health Perspectives* 124, no. 6 (October 26, 2015). doi:10.1289/ehp.1510037.
25. "Greenhouse Effects," *Wikipedia*, https://en.wikipedia.org/wiki/Greenhouse_effects.
26. "Ozone," *Wikipedia*, https://en.wikipedia.org/wiki/Ozone.
27. Marcia Bjornerud, *Reading the Rocks* (Cambridge, MA: Westview Press 2005). As she comments, the temperature of the surface of Venus is 806°F. The melting point of lead is 630°F; see "Lead," *Wikipedia: The Free Encyclopedia*, Wikimedia Foundation, January 10, 2018, accessed January 12, 2018, https://en.wikipedia.org/wiki/Lead.

28. R. A. Berner, "Atmospheric Oxygen over Phanerozoic Time," *Proceedings of the National Academy of Sciences* 96, no. 20 (September 28, 1999): 10955–57. doi:10.1073/pnas.96.20.10955. See Figure 2.
29. Abstracted from *The Wonder of Water.*
30. Stephan Harding and L. Margulis, "Water Gaia: 3.5 Thousand Million Years of Wetness on Planet Earth," in *Gaia in Turmoil: Climate Change, Biodepletion, and Earth Ethics in an Age of Crisis,* Eileen Crist and H. Bruce Rinker, eds. (Cambridge, Mass: MIT Press, 2010), 41–59. The authors defend a "Gaian view" of water's retention on Earth. They entitle their thesis: "Life Retained Planetary Water."
31. "Ozone," *Wikipedia,* https://en.wikipedia.org/wiki/Ozone.
32. "Electromagnetic Radiation," in *Encyclopaedia Britannica,* 15th edition (1994), vol. 18, p. 203. Emphasis added.
33. Fred Hoyle, "The Universe: Past and Present Reflections," *Engineering and Science* (1981): 8–12.
34. Ashwini Kumar Lal and Rhawn Joseph, "Big Bang? A Critical Review," *Journal of Cosmology* 6 (2010): 1533–1547.

4. The Gift of the Leaf

1. Charles Darwin, "To Frederick Watkins," letter of August 18, 1832 from Monte Video, Riv. Plata, *Darwin Correspondence Project,* University of Cambridge, https://www.darwinproject.ac.uk/letter/?docId=letters/DCP-LETT-181.xml.
2. "Maple," *Wikipedia: The Free Encyclopedia,* Wikimedia Foundation, January 7, 2018, accessed January 12, 2018, https://en.wikipedia.org/wiki/Maple.
3. John Roach, "Source of Half Earth's Oxygen Gets Little Credit," *National Geographic News,* June 2004, accessed January 12, 2018, *news.nationalgeographic.com/news/2004/06/0607_040607_phytoplankton.html.*
4. See Paul F. Ciesielski, "Transition of Plants to Land," *University of Florida,* via Wayback Machine, Archive.org, October 9, 1999, accessed January 12, 2018, https://web.archive.org/web/19991009125017/http://www.clas.ufl.edu/users/pciesiel/gly3150/plant.html. The oldest fossils reveal evolution of non-vascular plants by the middle to late Ordovician Period (~450–440 m.y.a.) on the basis of fossil spores. See also Hans Steur, "*Cooksonia,* a very old land plant (I)," *Hans' Paleobotany Pages,* May 10, 2017, accessed January 12, 2018, https://steurh.home.xs4all.nl/engcook/ecooks.html.
5. Ciesielski, op. cit.; Steur, op. cit.
6. R. W. Gess, "The earliest record of terrestrial animals in Gondwana: a scorpion from the Famennian (Late Devonian) Witpoort Formation of South Africa," *African Invertebrates* 54, no. 2 (2013): 373–379, doi:10.5733/afin.054.0206. The oldest unequivocal air-breathing organism is a myriapod, *Pneumodesmus newmani,* from the late Silurian (428 million years ago), described in Michael Hopkin, "Fossil Find Breaks Age Record," *Nature,* January 27, 2004, accessed January 12, 2018, http://www.nature.com/news/2004/040126/full/news040126-1.html. The first arachnids are described in Russell J. Garwood and Jason A. Dunlop, "Trigonotarbids," *Geology Today* 26, no. 1 (February 2010): 34–37. doi:10.1111/j.1365-2451.2010.00742.x.

7. Michael S. Engel and David A. Grimaldi, "New Light Shed on the Oldest Insect," *Nature* 427, no. 6975 (February 12, 2004): 627–30. doi:10.1038/nature02291.

8. J. N. Maina, "Structure, Function and Evolution of the Gas Exchangers: Comparative Perspectives," *Journal of Anatomy* 201, no. 4 (October 2002): 281–304.

9. Ibid., 284. Maina comments: "As a respiratory medium, air is a more cost-effective respiratory fluid: water is 50 times more viscous than air; the concentration of dissolved oxygen in water is about one-thirtieth that in air; the rate of diffusion of oxygen in water is lower by a factor of… [eight thousand]… All other conditions being equal, owing to the greater viscosity of water, compared with air breathing, water breathing requires more energy to procure an equivalent amount of oxygen."

10. George Wald, "The Origin of Life," *Scientific American* 191, no. 2 (August 1954), 44–53.

11. Ibid.

12. Scripps Institution of Oceanography, "Measuring Global Photosynthesis Rate: Earth's Plant Life 'Recycles' Carbon Dioxide Faster than Previously Estimated," *Science Daily*, October 2, 2011, accessed January 11, 2018, https://www.sciencedaily.com/releases/2011/09/110928222003.htm.

13. Ian M. Campbell, *Energy and the Atmosphere* (London: Wiley, 1977), 1.

14. Jeremy M. Berg, John L. Tymoczko, and Lubert Stryer, *Biochemistry*, 5th ed. (New York: W.H. Freeman, 2002), section 19.2: "What happens when light is absorbed by a molecule such as chlorophyll? The energy from the light excites an electron from its ground energy level to an excited energy level… This high-energy electron can have several fates. For most compounds that absorb light, the electron simply returns to the ground state and the absorbed energy is converted into heat. However, if a suitable electron acceptor is nearby, the excited electron can move from the initial molecule to the acceptor … This process results in the formation of a positive charge on the initial molecule (due to the loss of an electron) and a negative charge on the acceptor and is, hence, referred to as *photoinduced charge separation*. The site where the separational change occurs is called the *reaction center*. We shall see how the photosynthetic apparatus is arranged to make photoinduced charge separation extremely efficient. The electron, extracted from its initial site by absorption of light, can reduce other species to store the light energy in chemical forms." View online at NCBI: https://www.ncbi.nlm.nih.gov/books/NBK22535/.

15. "Thylakoid," *Wikipedia: The Free Encyclopedia*, The Wikimedia Foundation, January 11, 2018, accessed January 11, 2018, https://en.wikipedia.org/wiki/Thylakoid. See also "Chloroplast," *Wikipedia: The Free Encyclopedia*, The Wikimedia Foundation, January 11, 2018, accessed January 11, 2018, https://en.wikipedia.org/wiki/Chloroplast.

16. Ibid.

17. R. E. Blankenship, "The Evolution of Photosynthesis," *Plant Physiology* 154, no. 2 (2011): 434–438.

18. Katharina F. Ettwig, Daan R. Speth, Joachim Reimann, Ming L. Wu, Mike S. M. Jetten, and Jan T. Keltjens, "Bacterial Oxygen Production in the Dark," *Frontiers in Microbiology* 3 (2012). doi:10.3389/fmicb.2012.00273.

19. Bugra Turan, Jan-Philipp Becker, Félix Urbain, Friedhelm Finger, Uwe Rau, and Stefan Haas, "Upscaling of Integrated Photoelectrochemical Water-

Splitting Devices to Large Areas," *Nature Communications* 7 (September 7, 2016): 12681, doi:10.1038/ncomms12681; N. S. Wigginton, "Mimicking the Oxygen Evolution Center," *Science* 348, no. 6235 (May 8, 2015): 644–46, doi:10.1126/science.348.6235.644.

20. Nick Lane, *Oxygen: the Molecule That Made the World* (Oxford and New York: Oxford University Press, 2002), 145.

21. Daniel D. Bikle, "Vitamin D Metabolism, Mechanism of Action, and Clinical Applications," *Chemistry & Biology* 21, no. 3 (March 2014): 319–29. doi:10.1016/j.chembiol.2013.12.016.

22. Gregory T. Carroll, L. Devon Triplett, Alberto Moscatelli, Jeffrey T. Koberstein, and Nicholas J. Turro, "Photogeneration of Gelatinous Networks from Pre-Existing Polymers," *Journal of Applied Polymer Science* 122, no. 1 (October 5, 2011): 168–74, https://doi.org/10.1002/app.34133.

23. Howard E. Strassler, "The Physics of Light Curing and its Clinical Implications," *Compendium: of Continuing Education in Dentistry* 32, no. 6 (2001), AEGIS Dental Network, https://www.aegisdentalnetwork.com/cced/2011/08/the-physics-of-light-curing-and-its-clinical-implications.

24. Plants require copious amount of water. A large tree can transpire 500 liters of water per day, while, by comparison, a large animal like an elephant requires only 100 liters per day. In addition to water itself, the growth of plants requires a supply of the essential elements needed for cellular metabolism (sodium, potassium, calcium, magnesium, iron, etc.). The thriving of terrestrial plants also necessitates the existence of soils able to retain water in the substrate during periods of relative drought.

25. Water's fitness for terrestrial life was reviewed in *The Wonder of Water*, Chapter 1, and its fitness to climb the stems of plants and trunks of trees was described in Chapter 5. The sections below are abstracted from *The Wonder of Water* with minor changes. A reader familiar with that monograph can skip the text that follows.

26. Philip Ball, *H_2O: A Biography of Water* (London: Weidenfeld & Nicolson, 1999), 23–24.

27. Ibid., 26.

28. Ibid.

29. Ibid.

30. "Surface Tension," *Wikipedia: The Free Encyclopedia*, Wikimedia Foundation, January 12, 2018, accessed January 12, 2018, https://en.wikipedia.org/wiki/Surface_tension, see data table.

31. "Gallium," *Wikipedia: The Free Encyclopedia*, Wikimedia Foundation, January 11, 2018, accessed January 12, 2018, https://en.wikipedia.org/wiki/Gallium.

32. Gerard V. Middleton and Peter R. Wilcock, *Mechanics in the Earth and Environmental Sciences* (Cambridge and New York: Cambridge University Press, 1994), 84.

33. Nyle C. Brady and Raymond Weil, *Elements of the Nature and Properties of Soils*, 15th edition (Harlow, Essex: Pearson Education Limited, 2016), 22.

34. Ibid., 22.

35. Matti Leisola, Ossi Pastinen, and Douglas D. Axe, "Lignin—Designed Randomness," *BIO-Complexity* 2012 (2012).
36. Because of the importance of this seemingly esoteric fitness of water for photosynthesis in large trees and consequently the provision of wood, the essential fuel for the high-temperature fire needed for metallurgy, I have included the section below, which is abstracted with minor editorial changes from *Fire-Maker* and *The Wonder of Water.*
37. Steven Vogel, *The Life of a Leaf* (Chicago: University of Chicago Press, 2010), Chapter 6.
38. N. Michele Holbrook and Maciej A. Zwieniecki, "Transporting Water to the Tops of Trees," *Physics Today* 61 (2008): 76–77.
39. Holbrook and Zwieniecki.
40. Vogel (2010), Chapter 6, p. 93.
41. Holbrook and Zwieniecki.
42. Ibid.
43. Ibid.
44. Melvin T. Tyree, "The Tension Cohesion Theory of Sap Ascent: Current Controversies," *Journal of Experimental Botany* 48, no. 315 (1997): 1753–1765.
45. Vogel (2010), 101.
46. Ibid., 91. Emphasis added.
47. Holbrook and Zwieniecki.
48. See Matti Leisola, Ossi Pastinen, and Douglas D. Axe, "Lignin—Designed Randomness," *BIO-Complexity* 2012 (2012). Lignin is an essential component of all plant cell walls and provides the essential element of strength necessary for the construction of tall woody trees. Because it is highly resistant to enzymatic catalysis, its breakdown in the soil is slow, allowing the formation of humus, which retains water and minerals in the soil. This in turn promoted the growth of large trees and allowed the build-up of vast volumes of undigested vegetation in the Carboniferous swamps, ultimately providing the coal for the steam engines of the early industrial age. Without lignin, there would be no woody plants, no wood, no coal, no charcoal, no fire, no pottery, certainly no free iron, and probably no other metals or metallurgy.
49. James C. Forbes, *Plants in Agriculture* (Cambridge and New York: Cambridge University Press, 1992), Figure 4.18, page 100, and section 4.9.1, "Thermal injury and its avoidance"; see also Hans Lambers et al., *Plant Physiological Ecology*, 2nd ed. (New York: Springer, 2008), 225–235.

5. Fitness for Vision

1. Jonathan Lear, *Aristotle: The Desire to Understand* (Cambridge and New York: Cambridge University Press, 1988), 1.
2. "Animal Echolocation," *Wikipedia: The Free Encyclopedia*, Wikimedia Foundation, January 11, 2018, accessed January 12, 2018, https://en.wikipedia.org/wiki/Animal_echolocation.

3. Joanna Moorhead, "Seeing with Sound," *The Guardian, US*, January 27, 2007, accessed January 12, 2018, https://www.theguardian.com/lifeandstyle/2007/jan/27/familyandrelationships.family2.

4. Kenneth C. Catania, "A Nose for Touch," *The Scientist*, September 1, 2012, accessed January 12, 2018, http://www.the-scientist.com/?articles.view/articleNo/32505/title/A-Nose-for-Touch/. The author (from Vanderbilt University) elsewhere comments: "In total, a single star contains about 25,000 domed Eimer's organs, each one served by four or so myelinated nerve fibers and probably about as many unmyelinated fibers… This adds up to many times more than the total number of touch fibers (17,000) found in the human hand—yet the entire star is smaller than a human fingertip." See K. C. Catania and J. H. Kaas, "Areal and callosal connections in the somatosensory cortex of the star-nosed mole," *Somatosens Mot Res*, 18 (2001): 303–11.

5. Carl Sagan, *Cosmos* (New York: Ballantine Books, 2013), 102.

6. G. L. Walls, *The Vertebrate Eye and its Adaptive Radiation* (New York: Hafner Publishing Co., 1963). See also Dan-Eric Nilsson, "Eye Evolution and Its Functional Basis," *Visual Neuroscience* 30, no. 1–2 (March 2013): 5–20. doi:10.1017/S0952523813000035. See as well Michael Land, *Encylopaedia Britannica*, s.v. "Structure and Function of Photoreceptors," accessed February 13, 2018, https://www.britannica.com/science/photoreception/Structure-and-function-of-photoreceptors.

7. H. Tabandeh, G. M. Thompson, P. Heyworth, S. Dorey, A. J. Woods, and D. Lynch, "Water Content, Lens Hardness and Cataract Appearance," *Eye* 8, no. 1 (January 1994): 125–29. doi:10.1038/eye.1994.25.

8. The highest-acuity non-camera type biological eye on earth is that of the dragonfly, but as Michael Land points out, this has a resolving power fifteen times less than that of the human eye. See article by Land, *Encylopaedia Britannica*, https://www.britannica.com/science/photoreception/Structure-and-function-of-photoreceptors.

9. Yingbin Fu, "Phototransduction in Rods and Cones," in *Webvision: The Organization of the Retina and Visual System*, H. Kolb, E. Fernandez, and R. Nelson, eds. (Salt Lake City: University of Utah Health Sciences Center, 2016), 533–580. Available at https://www.ncbi.nlm.nih.gov/books/NBK11530/pdf/Bookshelf_NBK11530.pdf. See table 2.

10. Land, *Encylopaedia Britannica*, https://www.britannica.com/science/photoreception/Structure-and-function-of-photoreceptors.

11. A. W. Snyder, "Photoreceptor Optics: Theoretical Principles," in *Photoreceptor Optics*, A. W. Snyder and R. Menzel, eds. (NewYork: Springer-Verlag, 2013), 38–55.

12. O. Packer and D. R. Williams, "Light, the Retinal Image, and Photoreceptors," in *The Science of Color*, S. Shevell, ed. (Boston; Elsevier, 2003), 41–102. The authors comment on page 64: "Efficient photon usage is promoted by the ability of photoreceptors to funnel light through long photopigment-filled outer segments. This ability is a result of waveguide properties."

13. Philip Charles Nelson, Sarina Bromberg, Ann Hermundstad, and Jesse M. Kinder, *From Photon to Neuron: Light, Imaging, Vision* (Princeton, New Jersey: Princeton University Press, 2017); J. Chen and A. P. Sampath, "Structure and Function of Rod and Cone Photoreceptors in Schachat" (1918), in *Ryan's Retina*, P. Andrew and

Charles P. Wilkinson, eds., 6th ed. (Edinburgh; New York: Elsevier, 2017), e-copy available at: https://books.google.co.th/books?redir_esc=y&hl=th&id=YWWwDgAAQBAJ&q=rods#v=onepage&q=rods&f=false; Packer and Williams, "Light, The Retinal Image, and Photoreceptors"; Fu, "Phototransduction in Rods and Cones."

14. The photoreceptors in the octopus eye are about 60 microns long—see J. J. Wolken, "Retinal Structure. Mollusc Cephalopoda: Octopus, Sepia," *Journal of Biophysical and Biochemical Cytology* 4, no. 6 (1958): 835–838—and about 3 microns in diameter—inferred from C. M. Talbot and J. N. Marshall, "The Retinal Topography of Three Species of Coleoid Cephalopod: Significance for Perception of Polarized Light," *Philosophical Transactions of the Royal Society B: Biological Sciences* 366, no. 1565 (March 12, 2011): 724–33, https://doi.org/10.1098/rstb.2010.0254. Talbot and Marshall give a maximum density of octopus photoreceptors as 90,000 per sq mm = cross sectional area of each photoreceptor of about 10 sq microns and diameter of about 3 microns. See also T. Yamamoto et al., "Fine Structure of Octopus Retina," *Journal of Cell Biology* 25 (1965): 365–359, and Kim Newkirk and Sony Kuhn, "Looking an Octopus in the Eye," University of Tennessee, College of Veterinary Medicine, September 25, 2013, accessed January 12, 2018, https://www.vetmed.wisc.edu/pbs/dubielzig/pages/coplow/PowerPoints/Cops%20pdf/Friday/Looking%20an%20octopus%20in%20the%20eye%209-25-13..pdf.

15. Nelson et al. (2017), 211; Land, *Encylopaedia Britannica*, https://www.britannica.com/science/photoreception/Structure-and-function-of-photoreceptors. As Land comments: "Pinhole eyes, in which the size of the pigment aperture is reduced, have better resolution than pigment cup eyes. The most impressive pinhole eyes are found in the genus *Nautilus*, a member of a cephalopod group that has changed little since the Cambrian Period (about 542 million to 488 million years ago). These organisms have eyes that are large, about 10 mm (0.39 inch) across, with millions of photoreceptors. They also have muscles that move the eyes and pupils that can vary in diameter, from 0.4–2.8 mm (0.02–0.11 inch), with light intensity. These features all suggest an eye that should be comparable in performance to the eyes of other cephalopods, such as the genus *Octopus*. However, *because there is no lens and each photoreceptor must cover a wide angle of the field of view, the image in the Nautilus eye is of very poor resolution*. Even with the pupil at its smallest, each receptor views an angle of more than two degrees, compared with a few fractions of a degree in *Octopus*. In addition, because the pupil has to be small in order to achieve even a modest degree of resolution, the image produced in the *Nautilus* eye is extremely dim. Thus, a limitation of pinhole eyes is that any improvement in resolution is at the expense of sensitivity; this is not true of eyes that contain lenses."

16. H. Kolb, "Facts and Figures concerning the Human Retina," in *Webvision: The Organization of the Retina and Visual System*, H. Kolb, E. Fernandez, and R. Nelson, eds. (Salt Lake City: University of Utah Health Sciences Center, 2016), 1721–1732, see page 1726.

17. R. Schlar, "An Eagle's Eye: Quality of the Retinal Image," *Science* 176 (1972): 920–922; Ivan R. Schwab, Richard R. Dubielzig, and Charles Schobert, *Evolution's Witness: How Eyes Evolved* (New York: Oxford University Press, 2012), 195.

18. Land, *Encylopaedia Britannica*, https://www.britannica.com/science/photoreception/Structure-and-function-of-photoreceptors. Emphasis added.

19. Fu, "Phototransduction in Rods and Cones," 541; see also discussion in Packer et al., "Light, The Retinal Image, and Photoreceptors," 41–102.
20. R. Gunter, H. G. W. Harding, and W. S. Stiles, "Spectral Reflexion Factor of the Cat's Tapetum," *Nature* 168, no. 4268 (August 18, 1951): 293–94. See also Ron Milo and Rob Phillips, "How Many Rhodopsin Molecules Are in a Rod Cell?" *Cell Biology by the Numbers*, 2015, accessed January 12, 2018, http://book.bionumbers.org/how-many-rhodopsin-molecules-are-in-a-rod-cell/.
21. L. Stryer, "Vision: From Photon to Perception," *PNAS* 93, no. 2 (1996): 557–559.
22. A. Despopoulos and S. Silbernagl, *Color Atlas of Physiology* (New York: Thieme Medical Publ., 1991), 306.
23. Land, *Encylopaedia Britannica*, https://www.britannica.com/science/photoreception/Structure-and-function-of-photoreceptors. Emphasis added.
24. Vadim Arshavsky, "Like Night and Day: Rods and Cones Have Different Pigment Regeneration Pathways," *Neuron* 36, no. 1 (September 26, 2002): 1–3. "Sustained vision requires continuous regeneration of visual pigments… this issue of *Neuron* … [reports] a novel enzymatic pathway uniquely designed to keep up with the high demand for cone pigment regeneration in bright light and to preclude rods from utilizing chromophore produced in daylight, when rods are not very useful for vision." See also Jin-Shan Wang and Vladimir J. Kefalov, "The Cone-Specific Visual Cycle," *Progress in Retinal and Eye Research* 30, no. 2 (March 2011): 115–28. https://doi.org/10.1016/j.preteyeres.2010.11.001.
25. S. Hecht, C. Haig, and A. M. Chase, "The Influence of Light Adaptation on Subsequent Dark Adaptation of the Eye," *The Journal of General Physiology* 20 (1937): 831–50, see Fig. 2; Also see the animation at "Dark Adaptation," *University of Calgary*, accessed January 12, 2018, http://www.ucalgary.ca/pip369/mod3/brightness/darkadaption; And Jin-Shan Wang and Vladimir J. Kefalov, "The Cone-Specific Visual Cycle," *Progress in Retinal and Eye Research* 30, no. 2 (March 2011): 115–28, https://doi.org/10.1016/j.preteyeres.2010.11.001: "Following exposure to bright light, cones dark adapt within 3–4 minutes while rods take over 30 minutes to fully restore their sensitivity."
26. The Airy Disc is named after George Biddell Airy, who first provided an explanation for the phenomenon in G. B. Airy, "On the Diffraction of an Object-glass with Circular Aperture," *Transactions of the Cambridge Philosophical Society* 5 (1835): 283–291. Available at https://archive.org/stream/transactionsofca05camb#page/n305/mode/2up/search/airy.
27. This formula is given at many sites: "Limitations on Resolution and Contrast: The Airy Disk," *Edmund Optics*, 2018, accessed January 12, 2018, https://www.edmundoptics.com/resources/application-notes/imaging/limitations-on-resolution-and-contrast-the-airy-disk/; "Airy Disk," J. E. Greivenkamp, *Field Guide to Geometrical Optics* (Bellingham, WA: SPIE Press, 2004), accessed January 12, 2018, https://www.spie.org/publications/fg01_p88_airy_disk?SSO=1; "Airy disk," *Wikipedia: The Free Encyclopedia*, Wikimedia Foundation, November 26, 2017, accessed January 12, 2018, https://en.wikipedia.org/wiki/Airy_disk.
28. M. D. Levine, *Vision in Man and Machine* (New York: McGraw-Hill Book Company, 1985), 68.
29. Ibid.

30. Website for calculating Airy disc diameter at "f/#, NA, and spot size," *CalcTool*, accessed February 13, 2018, http://www.calctool.org/CALC/phys/optics/f_NA.
31. H. B. Barlow, "The Physical Limits of Visual Discrimination," in *Photophysiology*, vol. 2, ed. A. C. Giese (New York: Academic Press, 1964), 163–202, see page 193; see also "Minute and Second of Arc," *Wikipedia: The Free Encyclopedia*, The Wikimedia Foundation, May 10, 2018, accessed June 6, 2018, https://en.wikipedia.org/wiki/Minute_and_second_of_arc.
32. "Minute and Second of Arc," *Wikipedia*, https://en.wikipedia.org/wiki/Minute_and_second_of_arc.
33. "Angular Size and Similar Triangles," *NASA*, accessed January 12, 2018, https://www.nasa.gov/sites/default/files/files/YOSS_Act_9.pdf. Calculated from figure of 0.5 degrees as the diameter across the lunar disc.
34. Lilliput was one of the islands inhabited by miniature people in Jonathan Swift's *Gulliver's Travels*.
35. Despopoulos, and Silbernagl, *Color Atlas of Physiology*.
36. Harvey Lodish, ed., *Molecular Cell Biology*, 4th ed. (New York, NY: Freeman, 2002). E-copy available https://www.ncbi.nlm.nih.gov/books/NBK21475/. In section 19.4 the authors comment: "Virtually all eukaryotic cilia and flagella are remarkably similar in their organization, possessing a central bundle of microtubules, called the axoneme, in which nine outer doublet microtubules surround a central pair of singlet microtubules… Regardless of the organism or cell type, the axoneme is about 0.25 μm in diameter, but it varies greatly in length, from a few microns to more than 2 mm."
37. Fu, "Phototransduction in Rods and Cones," see Table 2.
38. Dan-Eric Nilsson, Eric J. Warrant, Sönke Johnsen, Roger Hanlon, and Nadav Shashar, "A Unique Advantage for Giant Eyes in Giant Squid," *Current Biology* 22, no. 8 (April 2012): 683–88. https://doi.org/10.1016/j.cub.2012.02.031. See also Mera McGrew, "The Biggest Eyes in the Animal Kingdom," *Mission Blue*, Sylvia Earle Alliance, August 20, 2012, accessed January 12, 2018, https://mission-blue.org/2012/08/the-biggest-eyes-in-the-animal-kingdom/. From this post: "'They are probably the largest eyes that have ever existed,' says Eric Warrant, a professor at the University of Lund in Sweden and an expert on animal vision."
39. S. Ritland, "The Allometry of the Vertebrate Eye," dissertation (University of Chicago, 1982), T28274. See also "Eye Size Chart," *Noadi's Art*, accessed January 12, 2018, http://www.noadi.net/EyeSizes.html.
40. From "Brobdingnag," *Wikipedia: The Free Encyclopedia*, Wikimedia Foundation, January 10, 2018, accessed January 12, 2018, https://en.wikipedia.org/wiki/Brobdingnag. "Unlike his account of Lilliput, Gulliver does not say exactly how big the people of Brobdingnag are. However, in at least two cases he states explicitly that a Brobdingnagian's eyes are 'above sixty feet' from the ground, giving a ratio of at least eleven to one. He also states that he would 'appear as inconsiderable to this nation as a Lilliputian would be among us', suggesting the same twelve to one ratio given for Lilliput was intended. Hailstones are almost 1,800 times as heavy as in Europe, consistent with the figure. Gulliver also describes visiting the chief temple in Lorbrulgrud, whose tower was the highest in the kingdom, but reports he 'came back disappointed, for the height is not above three thousand foot', which 'allowing for the

difference in size between those people and us in Europe' is 'not equal in proportion to Salisbury steeple.'"

41. Frits Warmolt Went, "The Size of Man," *American Scientist* 56, no. 4 (1968): 40.

42. Steven Vogel, *Comparative Biomechanics: Life's Physical World* (Princeton, NJ: Princeton University Press, 2003), 18.

43. J. Liang and D. R. Williams, "Aberrations and retinal image quality of the normal human eye," *Journal of the Optical Society of America A* 14, no. 11 (November 1997): 2873–2883.

44. Ibid. Liang and Williams comment: "The biological optics of the human eye reveal local ir- regularities that are not present in man-made optics. The wave-front measurements [discussed in this paper] provide the most complete description of the eye's aberrations, showing conclusively that for large pupils the eye suffers from higher-order, irregular aberrations. These irregular aberrations do not reduce visual performance when the pupil is small (3 mm), such as in very-high-light-level conditions."

45. Michael Denton, *Nature's Destiny: How the Laws of Biology Reveal Purpose in the Universe* (New York: Free Press, 1998), 65.

46. R. H. Smythe, *Vision in the Animal World* (New York: St Martin's Press, 1975), 94.

47. Denton, *Nature's Destiny*, 65. See also Amritlal Mandal, Mohammad Shahidullah, and Nicholas A. Delamere, "Hydrostatic Pressure-Induced Release of Stored Calcium in Cultured Rat Optic Nerve Head Astrocytes," *Investigative Ophthalmology & Visual Science* 51, no. 6 (June 2010): 3129–38. https://doi.org/10.1167/iovs.09-4614.

48. Patrick Moore, "Refracting Telescopes," in *Encyclopedia of Astronomy and Astrophysics* (Nature Publishing Group, 2001), available at http://www.astro.caltech.edu/~george/ay20/refracting-telescopes.pdf.

49. Wald, "Life and Light."

50. See George Wald's Nobel Lecture: George Wald, "The Molecular Basis of Visual Excitation," Nobel Lecture, December 12, 1967, available at http://www.ghuth.com/images/waldlecture.pdf.

51. "Planck constant," *The NIST Reference on Constants, Units, and Uncertainty*, National Institute of Standards and Technology, https://physics.nist.gov/cgi-bin/cuu/Value?h|search_for=universal_in!, accessed August 17, 2018. This resource gives the Planck constant as 6.626 070 040(81) x 10^{-34} J s.

52. D. R. Whikehart, *Biochemistry of the Eye* (Boston: Butterworth-Heinemann, 2003), 108.

53. G. L. Walls, *The Vertebrate Eye* (New York: Hafner Publishing Company, 1963), 652.

6. The Anthropocentric Thesis

1. Loren C. Eiseley, *The Immense Journey* (New York: Vintage Books, 1973), 161–162.

2. Hans Moravec, *Robot: Mere Machine to Transcendent Mind* (New York: Oxford University Press, 2000); Gaby Wood, *Living Dolls: A Magical History of the Quest*

for Mechanical Life (London: Faber, 2002); Nick Bostrom, *Superintelligence: Paths, Dangers, Strategies* (Oxford: Oxford University Press, 2014).

3. Bostrom, *Superintelligence*.

4. Carl Sagan, *Contact: A Novel* (New York: Pocket Books, 1997), 191.

5. Carl Sagan, *Cosmos* (New York: Random House, 2002), 23.

6. David M. Toomey, *Weird Life: The Search for Life That Is Very, Very Different from Our Own* (New York: Norton, 2013); Dirk Schulze-Makuch and David J. Darling, *We Are Not Alone: Why We Have Already Found Extraterrestrial Life* (Oxford: Oneworld, 2011); Marc Kaufman, *First Contact: Scientific Breakthroughs in the Hunt for Life beyond Earth*, 1st Simon & Schuster hardcover ed. (New York: Simon & Schuster, 2011); Jack Cohen and Ian Stewart, *Evolving the Alien* (London: Ebury Press, 2002). Such books have disseminated somewhat uncritically the notion that the laws of nature permit a vast zoo of alternative types of life based on completely different chemistries. Yet although the concept of a universe teeming with life on every exo-planet (some carbon-based and some exotic) holds great appeal, not every recent author is quite as confident that the cosmos teems with alien life far more advanced than ourselves; see David Waltham, *Lucky Planet: Why Earth Is Exceptional—and What That Means for Life in the Universe* (New York: Basic Books, 2014), and Paul Davies, *The Eerie Silence: Renewing Our Search for Alien Intelligence*, 1st U.S. ed. (Boston: Houghton Mifflin Harcourt, 2010). The notion that life on Earth is a gigantic fluke, unlikely to have been repeated anywhere else in the universe, was recently intoned by secular guru and astrophysicist Brian Cox on the BBC program, *The Human Universe*; see the episode, "Are We Alone?" aired on October 21, 2014.

7. Alfred Russel Wallace, *The World of Life: A Manifestation of Creative Power, Directive Mind and Ultimate Purpose* (London: Chapman and Hall, 1910); Lawrence Joseph Henderson, *The Fitness of the Environment: An Enquiry into the Biological Significance of the Properties of Matter* (New York: McMillan, 1913).

8. A. E. Needham, *Uniqueness of Biological Materials* (UK: Pergamon Press, 1965); George Wald, "Fitness in the universe: Choices and necessities," *Origins of Life* 5 (1974): 7–27; N. R. Pace, "The Universal Nature of Biochemistry," *Proceedings of the National Academy of Sciences USA* 98 (2001): 805–808; L. N. Irwin and D. Schulze-Makuch, *Cosmic Biology: How Life Could Evolve on Other Worlds*, Praxis 2011 edition (New York: Springer, published in association with Praxis Pub., 2010), 29; K. W. Plaxco and M. Gross, *Astrobiology: A Brief Introduction* (Baltimore: Johns Hopkins University Press, 2011), Chapter 1.

9. Carl Sagan, *Cosmos* (New York: Ballantine Books, 1985), 105.

10. Paul Davies, *Accidental Universe* (Cambridge, UK: Cambridge University Press, 1982); J. D. Barrow and F. J. Tipler, *Anthropic Cosmological Principle* (Oxford: Oxford University Press, 1988); J. R. Gribbin and M. J. Rees, *Cosmic Coincidences* (UK: Black Swan, 1991); Martin J. Rees, *Just Six Numbers: The Deep Forces That Shape the Universe* (New York: Basic Books, 2000).

11. T. H. Huxley, *Man's Place in Nature* (New York: Modern Library, 2001), 71. Available at https://archive.org/stream/evidenceastomans00huxl#page/70/mode/2up/search/question+.

12. William Bains and Dirk Schulze-Makuch, "The Cosmic Zoo: The (Near) Inevitability of the Evolution of Complex, Macroscopic Life," *Life* 6, no. 3 (June 30, 2016): 25. https://doi.org/10.3390/life6030025.
13. Ibid.
14. Rob Phillips, *Physical Biology of the Cell*, second edition (London and New York: Garland Science, 2013).
15. Michael J. Denton, *Evolution: Still a Theory in Crisis* (Seattle: Discovery Institute Press, 2016), 250.
16. Carl Sagan, "Encyclopaedia Galactica," PBS program *Cosmos*, Episode 12 (December 14, 1980), 01:24 minutes in.
17. Had the "tape of scientific discovery" played out differently, the unique fitness paradigm I am defending here could have long ago been invalidated and the phantasma of the Cantina bar and the many worlds doctrine justified. For it *might have turned out*, as chemical knowledge of the atoms of the periodic table were increasingly defined in the nineteenth century, that many atoms possessed or even exceeded the fitness of the carbon atom for the purposes of building complex chemical structures. In such a scenario, already by the mid-nineteenth century any idea that nature was uniquely fit for carbon-based life would have been robustly invalidated. Again, it *might have turned out* that water was just a typical fluid, and that many other fluids possessed a similar suite of fitness and could have just as easily served as the matrix of carbon-based life. It *might have turned out* that many complex material forms had come into existence which possessed an equivalent set of elements of fitness as the cell, or exceeded the fitness of the cell altogether. Again, it *might have turned out* that we would by now understand how sentience and mind arise from matter and had already built, by 2014, Ray Kurzweil's superintelligent sentient machines, far surpassing those of modern humans. It *might have been the case* that the origin of life was already solved, removing any need for proposals of as yet undetected elements of generative fitness in nature which drew the living cell from the laws of inanimate chemistry. The fact is that the basic constituents of life on Earth—carbon atom, water, the cell, and so forth—do appear to be absolutely uniquely fit for their biological roles, and that all of the long-standing enigmas—of the origin of life, of man's intellectual abilities, of the major bio-forms on earth, and of the nature and origin of mind and sentience—still pose an existential challenge to the Darwinian and mechanistic world view.
18. Freeman Dyson, "Energy in the Universe," *Scientific American* 225 (1971): 50–59. Quoted in Barrow and Tipler, *The Anthropic Cosmological Principle*, 318.

A. Doing Without Sunlight

1. Kevin W. Plaxco and Michael Gross, *Astrobiology: A Brief Introduction*, 2nd ed. (Baltimore: Johns Hopkins University Press, 2011), 202–203.
2. Ibid., 203.
3. "Hydrothermal vent," *Wikipedia: The Free Encyclopedia*, Wikimedia Foundation, February 11, 2018, accessed February 15, 2018, http://en.wikipedia.org/wiki/Hydrothermal_vent.

4. Zoran Minic and Guy Hervé, "Biochemical and Enzymological Aspects of the Symbiosis between the Deep-Sea Tubeworm Riftia Pachyptila and Its Bacterial Endosymbiont," *European Journal of Biochemistry* 271, no. 15 (July 14, 2004): 3093–3102; Gerard Muyzer and Alfons J. M. Stams, "The Ecology and Biotechnology of Sulphate-Reducing Bacteria," *Nature Reviews: Microbiology* 6, no. 6 (June 2008): 441–54. Internal references removed.
5. N. R. Pace, "A Molecular View of Microbial Diversity and the Biosphere," *Science* 276, no. 5313 (May 2, 1997): 734–40.
6. Minic and Hervé, "Biochemical and Enzymological Aspects."
7. "Hydrothermal vent," *Wikipedia*, https://en.wikipedia.org/wiki/Hydrothermal_vent.
8. Minic and Hervé, "Biochemical and Enzymological Aspects."
9. See Muyzer and Stams, "The Ecology and Biotechnology of Sulphate-Reducing Bacteria."
10. George Wald, "Light and Life," *Scientific American* 201, no. 4 (1959): 92–108.
11. Nick Lane, *The Vital Question: Why Is Life the Way It Is?* (London: Profile Books, 2015), Chapter 3.
12. C. Smith, "Chemosynthesis in the Deep-Sea: Life without the Sun," *Biogeosciences Discussions* 9, no. 12 (December 4, 2012): 17037–52. Internal references removed. Emphasis added.
13. Ibid. Internal references removed.
14. Wald, "Life and Light."
15. Pace, "A Molecular View of Microbial Diversity and the Biosphere." And the "exotic" does not stop with means of obtaining metabolic energy; it extends to the ability of many microbes to thrive in "extreme" environmental conditions that would kill us in instants. As Lynn Rothschild put it in a recent *Nature* paper describing the newly uncovered universe of so-called extremophiles [Lynn J. Rothschild and Rocco L. Mancinelli, "Life in Extreme Environments," *Nature* 409, no. 6823 (February 22, 2001): 1092–1101, doi:10.1038/35059215]: "'Extremes' include physical extremes (for example, temperature, radiation or pressure) and geochemical extremes (for example, desiccation, salinity, pH, oxygen species or redox potential)." It now seems that virtually every aqueous habitat on Earth where the temperature range is compatible with stability of the organic compounds, no matter how acidic or alkaline, no matter how toxic and alien to our familiar conceptions, is home to some exotic, extremophile bacterial species, just as practically every possible means of energy procurement is exploited by some exotic (exotic from our perspective) life form. As she points out: "[I]n the past few decades we have come to realize that where there is liquid water on Earth, virtually no matter what the physical conditions, there is life. What we previously thought of as insurmountable physical and chemical barriers to life, we now see as yet another niche harbouring 'extremophiles.'"
16. Nadia Drake, "Subterranean Worms from Hell," *Nature*, June 1, 2011; G. Borgonie, A. García-Moyano, D. Litthauer, W. Bert, A. Bester, E. van Heerden, C. Möller, M. Erasmus, and T. C. Onstott, "Nematoda from the Terrestrial Deep Subsurface of South Africa," *Nature* 474, no. 7349 (June 2, 2011): 79–82. G. Borgonie, B. Linage-Alvarez, A. O. Ojo, S. O. C. Mundle, L. B. Freese, C. Van Rooyen, O. Kuloyo, et

al., "Eukaryotic Opportunists Dominate the Deep-Subsurface Biosphere in South Africa," *Nature Communications* 6 (November 24, 2015): 8952. From the abstract of the latter paper: "We report on the discovery in deep-subsurface fissure biofilm of Protozoa, Fungi, Platyhelminthes, Rotifera, Annelida, Arthropoda and additional Nematoda. Calculations indicate that food availability, not O_2, is a limiting factor for population growth. Video footage shows several types of biofilms growing on the fissure rock face, and collection of that biofilm establishes that this is the site where the Eukarya reside. The discovery of a complex group of interacting Eukarya in the deep subsurface indicates the biosphere on Earth is larger than previously determined and is significant for the search for life on other planets, particularly the planet Mars." Additionally, see M. A. Lever, O. Rouxel, J. C. Alt, N. Shimizu, S. Ono, R. M. Coggon, W. C. Shanks, et al., "Evidence for Microbial Carbon and Sulfur Cycling in Deeply Buried Ridge Flank Basalt," *Science* 339, no. 6125 (March 15, 2013): 1305–8.

17. Catherine Brahic, "Huge Hidden Biomass Lives Deep Beneath the Oceans," *New Scientist*, May 22, 2008, accessed February 15, 2018, https://www.newscientist.com/article/dn13960-huge-hidden-biomass-lives-deep-beneath-the-oceans. See also Erwan G. Roussel, Marie-Anne Cambon Bonavita, Joël Querellou, Barry A. Cragg, Gordon Webster, Daniel Prieur, and R. John Parkes, "Extending the Sub-Sea-Floor Biosphere," *Science* (New York, N.Y.) 320, no. 5879 (May 23, 2008): 1046. https://doi.org/10.1126/science.1154545.
18. Brahic, "Huge Hidden Biomass."
19. Lynn J. Rothschild and Rocco L. Mancinelli, "Life in Extreme Environments," *Nature* 409, no. 6823 (February 22, 2001): 1092–1101. doi:10.1038/35059215.
20. Jennifer L. Eigenbrode, Roger E. Summons, Andrew Steele, Caroline Freissinet, Maëva Millan, Rafael Navarro-González, Brad Sutter, et al., "Organic Matter Preserved in 3-Billion-Year-Old Mudstones at Gale Crater, Mars," *Science* 360, no. 6393 (June 8, 2018): 1096–1101. https://doi.org/10.1126/science.aas9185.

B. Fermentation and Cellular Respiration

1. George Wald, "The Origin of Life," *Scientific American* 191, no. 2 (August 1954), 44–53.
2. Jeremy M. Berg, John L. Tymoczko, and Lubert Stryer, *Biochemistry*, 5th ed. (New York: W.H. Freeman, 2002), section 16; available online at: https://www.ncbi.nlm.nih.gov/books/NBK21154.
3. Redox state refers to the tendency of an atom or molecule to accept electrons or donate them, a high redox value indicating a high reducing potential (tendency to donate electrons) and a low redox value indicating a high oxidizing potential (tendency to accept electrons).
4. For example, see Berg et al., *Biochemistry*, section 18.
5. Zoran Minic and Guy Hervé, "Biochemical and Enzymological Aspects of the Symbiosis between the Deep-Sea Tubeworm Riftia Pachyptila and Its Bacterial Endosymbiont," *European Journal of Biochemistry* 271, no. 15 (July 14, 2004): 3093–3102; Gerard Muyzer and Alfons J. M. Stams, "The Ecology and Biotechnology

of Sulphate-Reducing Bacteria," *Nature Reviews: Microbiology* 6, no. 6 (June 2008): 441–54.

6. See Lynn Rothschild, "Life in Extreme Environments," *Ad Astra* 14, no. 1 (2002), National Space Society, accessed February 15, 2018, http://www.nss.org/adastra/volume14/rothschild.html.
7. N. R. Pace, "A Molecular View of Microbial Diversity and the Biosphere," *Science* 276, no. 5313 (May 2, 1997): 734–40.
8. G. E. Nilsson, *The Respiratory Physiology of Vertebrates* (Cambridge: Cambridge University Press, 2010), Chapter 9.
9. Ibid.
10. Ibid.
11. J. N. Maina, "Structure, Function and Evolution of the Gas Exchangers: Comparative Perspectives," *Journal of Anatomy* 201, no. 4 (October 2002): 281–304.
12. Lisa A. Levin, "Anaerobic Metazoans: No Longer an Oxymoron," *BMC Biology* 8, no. 1 (2010): 31.
13. For instance (see Pace, op. cit.), the diplomonad *Giardia* is an anaerobic parasite found in contaminated water that causes the gastrointestinal disease giardiasis. Some yeasts also are anaerobes, having no ability to utilize oxygen, and presumably evolved from oxygen-utilizing relatives.

Figure Credits

Chapter 1

Figure 1.1. Sunrise at Stonehenge. By Andrew Dunn (CC BY-SA 2.0 (https://creativecommons.org/licenses/by-sa/2.0). From Wikimedia Commons.

Chapter 2

Figure 2.1. The life-giving Sun. © Leonid Tit/Adobe Stock (stock.adobe.com).

Figure 2.2. The solar spectrum. By Nick84. CC BY-SA 3.0 (https://creativecommons.org/licenses/by-sa/3.0), via Wikimedia Commons. From John F. B. Mitchell, "The 'Greenhouse' Effect and Climate Change," *Reviews of Geophysics* 27, no. 1 (1989), 115–139.

Figure 2.3. The electromagnetic spectrum. By NASA and Philip Ronan, Gringer. (CC BY-SA 3.0 (https://creativecommons.org/licenses/by-sa/3.0). Public domain via Wikimedia Commons.

Figure 2.4. Interaction of EM radiation with matter. By Brian Gage and Michael Denton.

Figure 2.5. The visual region. By Brian Gage and Michael Denton.

Figure 2.6. Visual band and adjacent near IR band. By Brian Gage and Michael Denton.

Figure 2.7. The Milky Way. By Steve Jurvetson via Flickr. CC BY 2.0 (http://creativecommons.org/licenses/by/2.0 or http://creativecommons.org/licenses/by/2.0). Via Wikimedia Commons.

Figure 2.8. Results of the double slit experiment. By Belsazar (Provided with kind permission of Dr. Tonomura). GFDL (http://www.gnu.org/copyleft/fdl.html) or CC-BY-SA-3.0 (http://creativecommons.org/licenses/by-sa/3.0/)]. Via Wikimedia Commons.

Chapter 3

Figure 3.1. Sunlight reaching through the atmosphere. © magann/Adobe Stock (stock.adobe.com).

Figure 3.2. Absorbance of EM radiation. By NASA (original). SVG by Mysid. Public domain via Wikimedia Commons.

Figure 3.3. Penetration of EM radiation into the atmosphere before absorption. By Brian Gage and Michael Denton, based on information from NASA. At "The Electromagnetic Spectrum," Imagine the Universe, NASA, March 2013, accessed July 13, 2018. https://imagine.gsfc.nasa.gov/science/toolbox/emspectrum1.html. See also: Anthony Watts, "The Solar Radio Microwave Flux," *Watts Up with That?* May 14, 2009, accessed June 26, 2018, https://wattsupwiththat.com/2009/05/14/the-solar-radio-microwave-flux/.

Figure 3.4. EM radiation penetrating the atmosphere. By Brian Gage and Michael Denton.

Figure 3.5. Absorption of EM radiation by water. Kebes at English Wikipedia. CC BY-SA 3.0 (https://creativecommons.org/licenses/by-sa/3.0) or GFDL (http://www.gnu.org/copyleft/fdl.html)]. Via Wikimedia Commons.

Figure 3.6. The oxygen content of the Earth. By Sauerstoffgehalt-1000mj.svg: LordToran derivative work: WolfmanSF (Sauerstoffgehalt-1000mj.svg). Public domain via Wikimedia Commons.

Figure 3.7. The narrow windows in the EM that facilitate photosynthesis. By Brian Gage and Michael Denton.

Chapter 4

Figure 4.1. Tropical foliage. © Jonny McCullagh/Adobe Stock (stock.adobe.com).

Figure 4.2. *Bombus barbutellus*. By gailhampshire from Cradley, Malvern, U.K (Bombus [Psithyrus] barbutellus ? CC BY 2.0 (http://creativecommons.org/licenses/by/2.0)], Via Wikimedia Commons.

Figure 4.3. Chlorophyll is the most important light-harvesting pigment. By Yikrazuul (Own work). Public domain via Wikimedia Commons.

Figure 4.4. The chloroplast. By SuperManu (own work based on Chloroplaste-schema.gif). GFDL (http://www.gnu.org/copyleft/fdl.html or CC BY-SA 3.0 (https://creativecommons.org/licenses/by-sa/3.0)]. Via Wikimedia Commons.

Figure 4.5. Light reactions and dark reactions. By Daniel Mayer (mav) for the original image. Vector version by Yerpo (Own work). GFDL (http://www.gnu.org/copyleft/fdl.html) or CC BY-SA 4.0-3.0-2.5-2.0-1.0 (https://creativecommons.org/licenses/by-sa/4.0-3.0-2.5-2.0-1.0). Via Wikimedia Commons.

Chapter 5

Figure 5.1. © Antonioguillem /Adobe Stock (stock.adobe.com).

Figure 5.2. The human eye. By Rhcastilhos and Jmarchn. CC BY-SA 3.0 (https://creativecommons.org/licenses/by-sa/3.0). Via Wikimedia Commons.

Figure 5.3. Computer-generated diffraction pattern. By Dicklyon at English Wikipedia. Public domain via Wikimedia Commons.

Figure 5.4. Image of computer-generated Airy disc. By Sakurambo at English Wikipedia. Public domain via Wikimedia Commons.

Figure 5.5. Visual region for photosynthesis and high-acuity vision. By Brian Gage and Michael Denton.

Chapter 6

Figure 6.1. The Ancient of Days. William Blake. Public domain via Wikimedia Commons.

Figure 6.2. One voice in the cosmic fugue? Designed by Carl Sagan and Frank Drake. Artwork prepared by Linda Salzman Sagan. Photograph by NASA Ames Research Center (NASA-ARC) (Ames Pioneer 10). Public domain via Wikimedia Commons.

Figure 6.3. Digital message. By Arne Nordmann (norro) Own drawing, 2005. GFDL (http://www.gnu.org/copyleft/fdl.html), CC-BY-SA-3.0 (http://creativecommons.org/licenses/by-sa/3.0/) or CC BY-SA 2.5 (https://creativecommons.org/licenses/by-sa/2.5)]. Via Wikimedia Commons.

INDEX

Made in the USA
Middletown, DE
19 September 2018